Building Balance Within You

Abdelaziz Benhaida

Dedication

To My Parents, Wife, Brothers, and all the sons and daughters who deserve the great relationships we want for them.

Acknowledgment

This is my first non-academic book. Writing such a book is harder than I thought and more rewarding than I could have ever imagined. None of this will see the light without my parents, wife, family, and best friends.

I'm eternally grateful to my father, who took in an extra mouth to feed when he didn't have to. He taught me discipline, tough love, manners, respect, and so much more that has helped me succeed in life.

To my mother, without whom I truly would have no idea where I'd be had she not given me a roof over my head or became the father whom I desperately needed.

To my wife, for always being the person I could turn to during my dark and desperate years. She sustained me in ways that I never knew that I needed. To my brothers, sisters, daughter, and son, thank you for letting me know that you had nothing but great memories of me.

Finally, thanks are due to the talented publishing team whose patience and expertise is much appreciated.

I hope that these ideas will provide you with the inspiration needed to find out more and stimulate your thinking along new, creative lines of taking care of yourself and your family by using your time, health, and opportunities more wisely.

About the Author

Abdelaziz is a blogger, affiliate marketer, and a freelancer. He blogs for online marketing purposes to help people and small businesses live a boss-free life via digital marketing. He does some of his best thinking when he can't sleep in the middle of the night or when he's away from his computer; in those quiet moments, he sometimes practically *"writes"* an entire piece in his head before he actually sits down to type it out.

He has a Master's degree in Business after finishing his Bachelor's degree in English Linguistics, in Morocco, Agadir. He got the motivation to start this book after seeing after observing two things. First, he noticed people wasting hours and hours of their time to no outcome, scrolling through social media feeds, watching YouTube, playing video games, and keeping *"busy"* with stuff that doesn't move their life forward constructively in the way they want.

Second, people are thinking about their life and retirement only after the age of 45-50, as the majority don't use their youth in the right way, which should ideally be the case. Good health is important for young people as it sets the stage for well-being later in life. Many more engage in behaviors that are dangerous not only to their current state of health but that also put their health at risk for years to come.

Promoting healthy lifestyles and taking steps to better educate and protect this group from health risks will ensure more productive lives.

Finally, the most important question: Why are we wasting opportunity? For the first time in human history, we can live full and meaningful lives, and it is this that, in my opinion, we squander. It's real; opportunity is within our grasp.

Do you sometimes feel overwhelmed by multiple projects, trapped by your own ideas? That you've taken on too much, working on ten different things without finishing any? These are the three points that motivated me to start writing this book.

Preface

This book is an intervention, one that we all need. As technology has permeated every aspect of our life, we have not had the chance to adapt to it socially. This has led to our society developing on its own, unconsciously and unguided. It is in this backdrop that I have decided to write this book.

I see how distracted we all have become. I'm sure you have felt it, too. In this fast-paced, ever-changing world, we have been unable to keep up. As a result, we have become twitchy, always anticipating the next change. This has made us perpetually distracted, unable to focus. We need to regain control of not only our bodies but also our mind.

This book is about enabling you to achieve precisely that. Based on my own personal journey, as well as my professional expertise, I have devised an effective strategy that helps me focus on my work. This is about my mind, and I know I need it to be under *my* control in order for me to reach my dreams. Using these tactics, I have managed to become everything I have ever wished, and now through this book, I hope you will be able to do the same, too!

Contents

Page Left Blank Intentionally

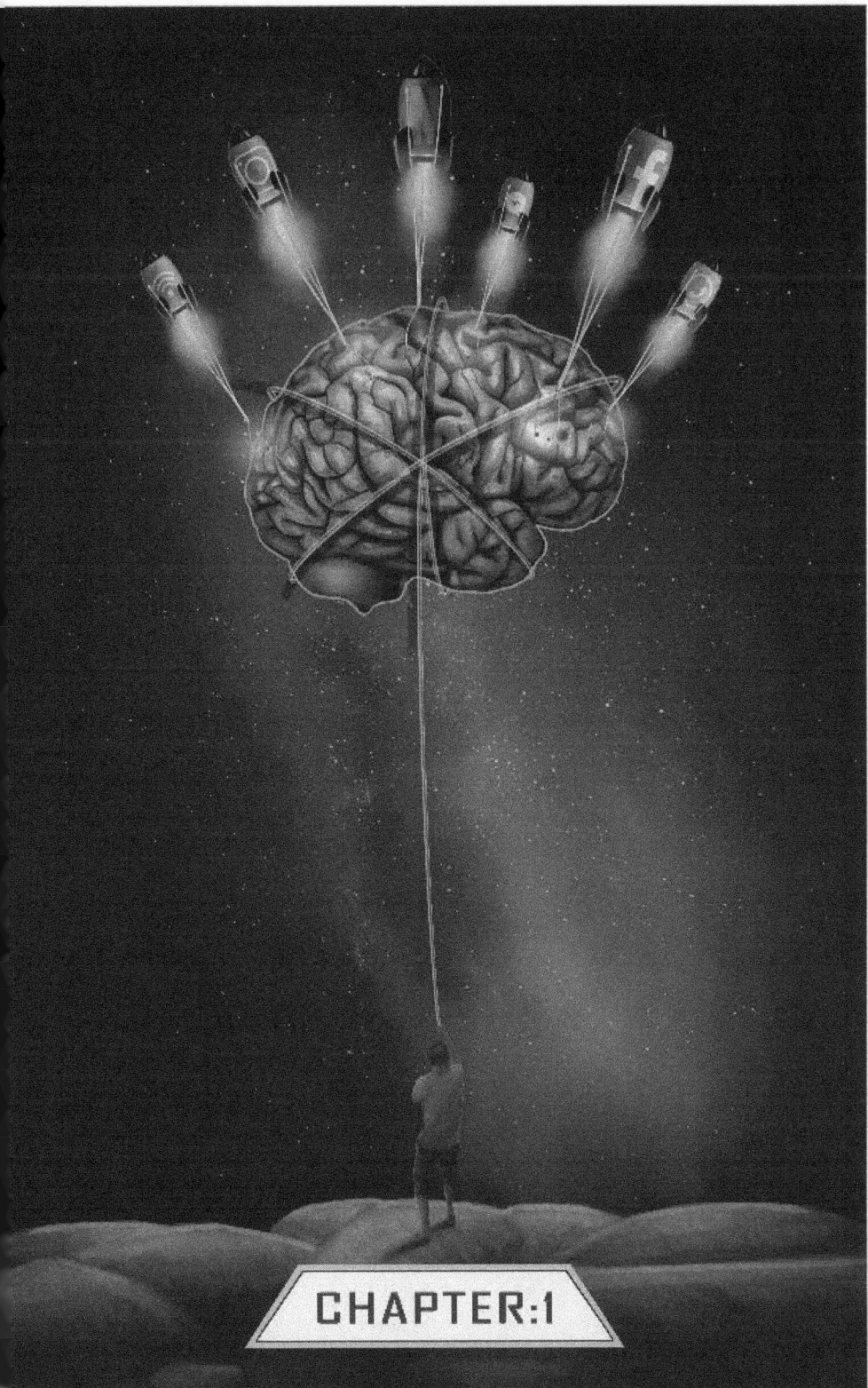

CHAPTER:1

Chapter 1
Introduction

"The pessimist sees difficulty in every opportunity. The optimist sees opportunity in every difficulty."

-Winston Churchill

You know the real problem is waves of distractions - late-night partying on a Tuesday night, texting under the table during a meeting, an overload of me aningless interactions, and feelings of frustration at trying to be perfect at almost everything. These are the fundamental problems that prevent most of us from achieving focus and some of life's most straightforward goals.

Whether it's a workplace, a social gathering, or only you as an individual, there are so many distractions that weigh heavily on your time and attention, driving you to distraction and failure. Here is a fun fact. Your brain is an unbelievable information processing machine and the most complex system ever acknowledged by human beings. It allows you to perform and accomplish significant tasks from discovering Mars and planning on traversing its landscape to painting Mona Lisa, and from building space shuttles to composing melodies. Still, you waste three hours of each day on Facebook, going through random people's pictures and profiles. Just think about what could have been accomplished or achieved during that time! Surprisingly, the answers to many of the struggles and problems you face each day that prevent you from becoming productive and successful, such as eating healthy, making the most of your time, availing the best opportunity at the right time, and finding peace, lies within you. We all know the answer to this already, yet we overlook what benefits the most and end up picking the thing that provides temporary pleasure.

"Distractions destroy action. If it's not moving you towards your purpose, leave it alone."

-Jermaine Riley

Distractions are the root of all failures. What can help you minimize distractions and remove failure from your dictionary? The ability to focus. But how can that work? See, your ability to focus on life's most challenging and meaningful tasks allows you to produce desired results in ways you haven't done before. It will enable you to ease things and find peace in life. It allows you to make things simple and focus on less. It helps you focus on the things that are more important in the grander scheme of things. Here, it doesn't mean 'less is more' but 'less is good.' It makes sense that by focusing on *less* at a time, you can ensure all of your focus and attention are paid to the tasks at hand. Focusing first on the smaller things will help you become more effective at different stages of your life.

Focusing on *less* will allow you to do less, which further allows you to have more time on your hand for important things. It will help you choose and prioritize, and in that process, eliminate the excuses that lead to problems, both as an individual and a professional.

Everywhere, people experience an epidemic of overwhelming distractions at work, even in their personal life. The trouble, as explained above, is the result of two things. Firstly, our brain may not be accustomed to the amount of information that it processes daily. For instance, did you know that the New York Times you read on Sunday contains more information than what the average 18th-century French nobleman learned in his lifetime? Secondly, the modern world is home to all the new technologies that are very good at distracting you.

Humans do not know how to manage time in the best manner possible while making the most out of new gadgets and technologies. Our human predecessors did not have to worry about this, considering technology wasn't as advanced – there were no smartphones, Google, or social media.

Why do we find it difficult to minimize distractions and manage time effectively? The challenge is that we haven't yet realized the real cost of distractions. They use up what is a limited supply of attention each day, making us far less effective if we need to do more in-depth thinking work.

An Era of Distraction

"Old robots are becoming more human, and young humans are becoming more like robots."

-Lorin Morgan-Richards, The Goodbye Family Unveiled

Humans are complicated, and we live in an awkward age. It is known as the era of information, but you can also refer to it as the era of distraction. All the technological inventions that surround you are trying to grab your attention. The tendency to rely on technology prevents most of you from realizing your real potential. As you immerse yourself in technology, constantly ringing telephones, and buzzing notifications, you inadvertently lose interest in things that matter in life. You spend hours and hours on the bed watching TV shows and movies on Netflix or scrolling through your social media feed rather than heading outside for a jog, sharing lunch with a co-worker, or spending time with family. Even more, you are connected with a flow of technology and a haze of multitasking.

When you're busy at the office, you face distractions from every direction; the computer in front of you, a constantly buzzing phone, email notifications, and loud co-workers. Let's not forget the luring call from the browser in front of you that contains an infinite amount of information that can lead you to a Bermuda Triangle from which it's impossible to escape, all the opportunities for online shopping, talking to long-lost friends, and an array of photos and videos that keep you hooked for hours. All this time, you have important deadlines to meet, and important business emails to respond to. As a result, not only are you wasting time that you should be using to achieve goals, but you are also becoming unhealthy.

Furthermore, you are facing difficulties in evaluating opportunities, establishing and maintaining human connections, and implementing simplicity and adopting a positive attitude. All of this is unparalleled and quite shocking. Evan Stutter beautifully illustrates the impact of technology on our lives, our health, and our vibes in his book, and how we can prevent that from happening.

"We all need a technological detox; we need to throw away our phones and computers instead of using them as our pseudo-defense system for anything that comes our way. We need to be bored and not have anything to use to shield the boredom away from us. We need to be lonely and see what it is we really feel when we are. If we continue to distract ourselves so we never have to face the realities in front of us, when the time comes and you are faced with something bigger than what your phone, food, or friends can fix, you will be in big trouble."

-Evan Sutter, Solitude: How Doing Nothing Can Change the World

The truth is, we have entered into this era without realizing its consequences. Of course, we knew that the Internet was increasing and how curious and excited we were about it. We also knew that mobile devices were gaining more and more recognition, and while some people resisted this invention, others welcomed it with open arms. Although the benefits offered by this opportunity were enticing, with the consistent distractions, the desperate call for our attention, the pressure of multitasking at the most basic level, and the invasion of our freedom, maybe we didn't realize how much of an impact this would have on our lives and our health. I believe, with so many things begging our attention, it's time we pay some attention to this problem.

Don't Let It Become an Addiction

"An over-indulgence of anything, even something as pure as water, can intoxicate."

-Criss Jami, Venus in Arms

If you did not know this before, there are some positive results of specific activities such as checking emails, browsing the internet, and checking your Twitter and Facebook every 10 minutes. This is why most of us become addicted to create connections and feel distracted when the relationship is not reciprocated. Various addictive activities, such as drugs or obsessive eating, have similar instant and positive results. You perform the activity, and you're instantly rewarded with something desirable, and you don't feel the impact of the negative consequences. I remember when I was in college, I spent a decent chunk of my adult life feeling directionless. I was juggling between two part-time jobs and had been a full-time student pursuing a lucrative career in my home country, Morocco, especially in Agadir. Back in the days, I was passionate about writing. I would construct complex, overlapping plotlines in my head about different topics, mainly about things that affected our lives. It turned out that I was only seeking instant results and positive feedback from editors, and when the expected feedback was not received, I felt demotivated and frustrated.

Later in life, when I began my quest for purpose and focus, I realized that the only reason I felt so frustrated was that I was seeking instant positive results. I failed to work on my weaknesses and immediately gave up without even trying and without finding the real reason behind the failure. Now you see, this is what happens when you're not connected with your inner self when you don't identify the principles you value the most in your life and feel distracted by material benefits. Once I felt I had reached the optimum destination in my life, I started viewing things from a completely different perspective.

I went on to achieve my master's from Ibn Zohr University in Morocco. While I learned an astonishing amount of material in the classroom, my time at university unexpectedly taught me more than I bargained for in life. This was all thanks to the many experiences I had and the people I met throughout those five years. I can thankfully say that my degree came with much more than just a scholarly education. I gained a lot of wisdom during those years.

They say the best chefs are the ones who can make gold from something simple. This is pretty much the slogan for me as a kitchen student, where I had to be smart and creative to make tasty meals out of tajine and couscous. There is no excuse for having a bad meal; you need to have the right knowledge and effort. Also, I learned a lot about using time correctly, though I figured this treasure in my final year of university. It was perhaps the most useful skill I developed as a part-time student.

How do you think Mark Zuckerberg, Bill Gates, Jeff Bezos, and Steve Jobs made it this far in their careers? They believed in themselves. They believed in the power of their dreams and continuously worked on improving their skills by focusing on the right elements. Let's assume you wake up on the right side of the bed one day and decide that you no longer want to be a part of the era of distraction. Do you think you can drop out that easily? Well, in a way, it is possible, but then you would be standing up against an entire culture that has high expectations from you.

Staying connected all the time and living up to the unrealistic expectations that society has of you can be very draining, whereas standing up against these expectations can be very challenging for some people. I'm not asking you to become rebellious against society and go off living alone on a remote island. But, you should try to rethink from different angles so that the system is more hospitable for you and not the other way around.

Building Balance within You

"We are what we repeatedly do. Excellence, then, is not an act, but a habit."

-Aristotle

In today's fast-paced world of pressure and deadlines, one finds it extremely difficult to find a balance between external and internal factors. Whether it's work, family, personal life, or something you firmly believe in, you are continuously struggling to find a balance between your needs and expectations.

Believe me when I say balance is an issue for almost everyone. There are so many things that you starve to balance in your life - work, time, health, home, leisure, personal experience, social, political, and emotional consciousness, religious and personal beliefs, and daily activities.

Finding balance in your life is very crucial, yet it is so hard that there are hundreds of articles, essays, and books written on it. If you merely google 'How to find balance within you,' you will find up to 333,000,000 results and why you need it so badly. Nevertheless, the question of why we are still unable to find it persists?

So How Do You Find Balance?

Well, the factors that require balance in our lives can be divided into two categories - internal and external. The reason why people are unable to find balance in life is that they are focusing on one element than the other. For instance, you may see yourself focusing on external factors such as bills, work, deadlines, and relationships. In contrast, you may be paying very little attention to what is troubling you internally.

On the other hand, you may find yourself focusing too much on the inner self-consciousness that you completely ignore all of life's experiences and treasures. Here is a short definition of internal and external factors that will help you understand the benefits of both sides of the poles.

Internal Factors

- Mind: Pushing yourself to face intellectual vs. availing opportunities that give you peace of mind

- Heart: Accepting love vs. giving love

- Health: Staying fit, exercising, and eating healthy vs. treating yourself to some delicious delicacies

External Factors

- Work: Pressurizing yourself to achieve deadlines vs. enjoying the experience and learning

- Social Life: Satisfying your social needs vs. pampering yourself

- Family and Friends: Pushing yourself to catch up with family and friends vs. defining healthy boundaries for all

- Leisure: Spending your time doing things you love vs. not overdoing it to the point that it becomes a routine

You can see that both ends are quite positive, but if you consider either end to an extreme, then something that is supposed to be positive might end up becoming very disastrous and detrimental. However, if you feel uncomfortable being pulled into any one of the directions, here are a few steps you can take to stop that from happening.

Recognize

Sit back and take a good look at your life, your choices, and the decisions you make daily and how they make you feel. Be honest with yourself and focus on the areas that you have been neglecting the most.

Inspect

Evaluate if you're bending more toward an external or internal end, or if there are any areas from each side that you would like to amplify.

Set Goals

Look at the factors above to help you eliminate the unnecessary clutter from your life and help you focus more on the necessary things. Decide what kind of focus you want to let into your life, and then set goals to reach that ultimate destination.

Action Plan

So you made it this far? Great! Now create an action plan that will lead you to your ultimate destination. Prepare a list of your daily, weekly, and monthly tasks that you need to achieve. Ask yourself, *how did I do it previously? Did it work?* If not, *what can you do differently that might lead to alternative results?* Do not, I repeat, do not neglect this step. Once you create an action plan, view it daily to keep track of your progress.

Speculate

How long did it take you to accomplish your goals? How focused were you? Were you able to handle your anxiety, stress, doubts, and pessimistic views? How does it feel to accomplish something you've wanted to achieve for a long time? Ask yourself these questions all the time. They will not only track your productivity but also push you in the right direction.

Prepare to Act

What are the inner thoughts that prevent you from focusing on your plan - fears, doubts, negativity, or weaknesses? Identify the things you say to get yourself off the track, such as 'Just one more drink, I promise, I'll stop from tomorrow.' Make a list and go through it. Now that you know what's stopping you from accomplishing the goal, you can easily overcome the hurdles.

Motivate

What are the things you would like to hear in those times of trouble? Imagine a friend going through the same situation. What would you say to motivate them? Be kind to yourself and give yourself the motivational talk you deserve. The balance will be no good if you're used to demeaning yourself.

Associate

Now that you've made it so far, look for a reason, a person, or a quote that helps you keep going in hard times. Honestly, I would recommend connecting with a person that you trust with your feelings and thoughts. This person can either be a family member, a close friend, or a spouse. Seek someone who will help you challenge your inner monsters and be the first one to clap for your smallest accomplishments.

Execute

To reach an ultimate state of mind or the desired destination, you need to input both time and effort to overcome your standard pattern of habits and create new ones. Trust me; if you continue following this path for the next three months, there is a 100% chance that you can enjoy a balanced and focused life without any distraction.

What to Expect from This Book?

This book is not a detailed memoir of historical events or cliché advice, but merely an elaborate disquisition of your current life with a handful of techniques to make it through life without any regrets. It is short, precise, and straightforward. In this book, we'll talk about some of the problems that we encounter on a daily basis as we try to live and find our meaning in this overwhelming world of technology and distractions. We'll look at some easy ways to overcome these problems, too.

As you open your mind to the next three topics - *health, opportunity*, and *time,* you will discover and explore the tools and desires needed to heal and rebuild your inner self. You will no longer succumb to your internal conflicts. Also, you will be able to focus on what matters the most in life and find it a lot easier to change your conventional ways of looking at life, yourself, and the things around you.

Building balance within you will make you capable of being independent and more productive in life. It will help you magnify your inner strengths and overcome your shortcomings. Once achieved, you can charge your batteries through it. Whatever is your current situation, whether you're struggling to gain a promotion, avoiding your problems, or finding an escape from reality, I assure you that your struggles do not define you. You can always recover old habits with new habits of focus, effectiveness, concentration, happiness, and meaningful relationships.

Readers, with genuine consideration, I urge you to focus on your priorities. Open the doors of growth and complete change as you read this book.

One Advice: Be patient. Self-growth is slow and vulnerable. There is no more significant investment than investing in yourself and receiving a motivated and highly inspired individual in both your personal and professional life.

It's obvious that you won't see the results immediately, but talking from past experiences, I assure you that you will observe benefits that will be encouraging and will improve your productivity at the workplace.

Questions to Ask Yourself

As we end chapter 1, here is a quick exercise that will help you evaluate what you learned so far. Ask yourself:

- How many times did you feel distracted while reading this chapter?

- How many times did you check your email while reading this chapter?

- How many things were noisy or distracting?

- How many people wanted to divert your attention?

In a not-so-real world, the answer to all these questions would be 'none' or 'zero.' I'm sure most of you had a lot of distractions coming your way from all sides, and I hope the answers to these questions will prove to be enlightening.

CHAPTER:2

Chapter 2
Health

"People shouldn't look at me and think life is one big piece of glamour. That's the marketing, the spin. Life is challenging. But I have courage, strength, and enough good health to see the positive."

-Carmen Dell' Orefice

As discussed in the first chapter, we live in an era of information, or simply put, the era of distraction. Technology has interfered in almost every sphere of our life. Some may agree that it has even come to control each day and minute of our life. No doubt, technology has made our lives convenient, fast-paced, and more comfortable.

It has also opened new panoramas for most of us. No one knows or is brave enough to make a risky guess about the next life-changing technological innovation. In this technology-dominated and fast-moving world, we deliberately need to take a step back, pause, and wonder about the lifestyle decisions we make, and the long-term effects of those decisions.

I'm sure a few of you would contradict the need for a healthy lifestyle. You're young, and retirement is decades away, so there is plenty of time till you can take care of your health. This is the way most youngsters look at life nowadays. No one can be blamed for this thought process because the majority of people prefer to stay at home as technology makes their lives easy. You believe that you have enough time to fix your life and bring about a change to live a comfortable retirement. Don't you?

To all those between the ages of 14 and 40 who share similar thinking, I urge you to start focusing on your health rather than ignoring it until you are forced to pay attention. You need to stop coasting through irregular sleeping schedules, skipping meals, binge eating, and staying locked inside of your room.

Because, if you start working on your health right now, it will eventually pay off during your retirement and you can avoid all the drawbacks that come with ailing health later in life. Alternatively, if you choose to delay it to your fifties, backaches, difficulty walking or running, and other medical issues are inevitable. Most retirees now realize that they should have managed their health when they were young.

If you want to avoid these problems and live a comfortable retirement when you can run, jog, and be active, start focusing on your health today. You will inevitably lose strength in old age, so prepare for this inescapable future by making an effort in the present. I mean, we all may fall sick for a short period at the very end of our life, but you don't want to spend the last seven to 10 years of your life debilitated due to chronic illness.

What people don't realize is that almost all chronic illnesses are caused by diet and lifestyle. So if you catch it in time, you probably can reverse it. You can likely reverse it even in the case of type 2 diabetes, where you may have lost the feeling in your toes. There's a lot of reversal that can potentially take place and allow you to reclaim your quality of life, but it's going to require that you make some serious changes right now. And to do that, you need to understand what true health is.

What is Health to You?

Health means you are going to wake up feeling rested and well aside from the odd flu and sickness that happens once in a while. For the most part, you're waking up feeling fresh, you're playing with your kids and with your grandkids actively, and you're roughhousing with them running around the backyard. You can enjoy your life to the fullest without chest pain and without feeling like you don't like what you see in the mirror. There is no pain in your body, and the true health comes without the need to take prescription medications that only give terrible side effects and generally end up making you feel even worse. In short, life is vibrant, hale, and healthy. You can do something now to either reverse the damage or prevent it altogether. To reclaim your health, you need to understand that what everybody else is doing, the *normal* isn't going to work. It's not going to allow you to maintain the life you want and the health you want to have up till the end.

You need to come to terms with the fact that the standard definition of lifestyle that is currently portrayed may look good in your 30s, or maybe even as old as your 40s or 50s, but it doesn't last, and you're eventually going to wish that you might have made some changes earlier in life. So make them now. I know that changing your entire diet and your lifestyle can sound intimidating, but if you try harder step by step, each day with small resolutions, you can attain good health that you enjoyed in your 20s.

Don't sell yourself short here; I understand it's hard, but eventually, you get used to your new lifestyle and your new diet - one that leads to feeling the best you've ever felt in your entire life. And all those things that you think you're going to have to deprive yourself of, it won't feel like deprivation after just a couple of short months.

It's hard to switch your habits and give up the things you're addicted to, but once you do it, it's all worth it. If you're conscientious enough to read this book, then you're in a place where you're doing it right. We're here building a community of like-minded people, so I encourage you to continue reading this book to see what it is to live a healthy and happy life.

Factors That Influence Your Health

"He who has health has hope, and he who has hope has everything."

-Arabian Proverb

Many factors influence your health. They are also called elements of health. One of the elements is what lies in our genes and our biology, while another comprise our behavior. This could involve whether you smoke, exercise, eat healthily, or sleep on time. Most people assume that their health is an outcome of their genes, their behavior, or attitude, or how frequently they get sick. But, honestly, it's not only your living pattern that governs how healthy you are. The social and physical environment surrounding you also plays an integral part in your lifestyle. These are referred to as social elements of health.

Road to Good Health

The social elements of health are the conditions you live, learn, work, play, and communicate. These elements can affect the health and well-being of not only you but also the community you live within. It includes things such as education standard, exposure to violence, the infrastructure of your community, or the ease of access to healthcare. These factors have a significant impact on your ability to participate in healthy activities, which ultimately affects your overall health and lifestyle. Some of the major social factors are discussed below.

Education Level

Like it or not, your level of education can have a substantial impact on how healthy you are. Education provides you with the tools you need to make good choices about your health. People with higher education levels are more likely to live healthier. They are more likely to take part in physical activities like exercising and running errands. Conversely, they are least expected to participate in unhealthy activities like smoking, taking drugs, or consuming alcohol. Education also leads individuals to higher-paid jobs. These jobs often come with benefits such as life insurance, healthcare facilities, healthy working conditions, and opportunities to connect with more people. All these contribute to a healthy and happy lifestyle if utilized the right way and at the right time.

Cleanliness

The area you live in has an impact on your lifestyle and health. People who are constantly exposed to unhealthy and poor conditions have greater chances of catching health problems. Pests, molds, malaria, violence, street crimes, and infectious diseases are some of the common problems people face when living in a dirty neighborhood. In order to maintain a healthy lifestyle, you need to ensure that your home, your area, and your neighborhood is free from risky health conditions like these. A community that is free from violence, crime, and pollution can provide children, adults, and elders with a safe, clean, and healthy place to carry out their physical activities.

If your house has access to natural and organic food, it becomes a lot easier to maintain a healthy lifestyle by eating a healthy diet. A flourishing neighborhood also provides employment, education, and transportation to its people. When all these blessings surround you, you automatically observe positive changes in your life.

Easy Access to Healthcare

How easily you can access healthcare is a massive element of your health. If you're provided with free healthcare, you are more likely to visit your doctor regularly. These visits can include general checkups, screening, and preventive measures to keep you from becoming a victim of chronic diseases. Unfortunately, not everyone has access to free healthcare or insurance to pay for their visits.

A few people don't even have enough to pay for the necessary transportation to go to a physician. These are just a few elements of health that make their way to the masses. Several other factors play an integral part of our lifestyle. Some of these are culture, gender, family, sexual identification, social support, social status, and financial status. Most of these factors have now become a deep-rooted part of our life that is hard to get rid of. If you feel there is a need to change them, now is the time to understand how they affect you, and only then can you take the necessary steps.

You know that healthy habits, such as eating well, exercising, and avoiding harmful substances, make sense, but did you ever stop to think about why you practice them? A healthy habit is any behavior that benefits your physical, mental, and emotional health. These habits improve your overall well-being and make you feel good.

Healthy habits are hard to develop and often require changing your mindset. Still, if you're willing to make sacrifices to better your health, the impact can be far-reaching, regardless of your age, sex, or physical ability…

If you give them a try, you may feel happier and more positive and might be able to get the most from life.

- **Connect** – connect with the people around you: your family, friends, colleagues, and neighbors. Spend time developing these relationships.

- **Be active** – you don't have to go to the gym. Take a walk, go cycling, or play a game of football. Find an activity that you enjoy and make it a part of your life. Learn more about getting active for mental wellbeing.

- **Keep learning** – learning new skills can give you a sense of achievement and new confidence.

- **Give to others** – even the smallest act can count, whether it's a smile, a thank you, or a kind word. Larger acts, such as volunteering at your local community center, can improve your wellbeing and help you build new social networks.

- **Be mindful** – be more aware of the present moment, including your thoughts and feelings, your body, and the world around you. Some people call this awareness *"mindfulness."* It can positively change the way you feel about life and how you approach challenges.

A healthy lifestyle leaves you fit, energetic, and at reduced risk for disease, based on the choices you make about your daily habits. Proper nutrition, regular exercise, and adequate sleep are the foundations for continuing good health. Managing stress in positive ways, instead of through smoking or drinking alcohol, reduces wear and tear on your body at the hormonal level. For a longer and more comfortable life, put together your plan for a healthy lifestyle, and live up to it.

Steps to Take

"Every single journey of your life starts with a healthy mind and a healthy journey."

-Jitendra Attra, Chakravyuh The Land of the Paharias

Hoping to change your lifestyle overnight is more delusional than you think. No human has ever been able to become healthy, eat healthily, reduce weight, or improve their sleep pattern in just one night. However, you can start taking baby steps, which will help you reach the end goal of a healthy life.

If you want to reduce your chances of infection or simply improve your overall life, just start your day by following these basic chores that contribute to personal hygiene.

Waking up Early

Getting up early gives you a kick start for the day ahead. Besides allowing you more hours for your work, it also boosts your speed. Studies have suggested that when a person gets up early, he is more energetic and takes lesser time to do a task that would take more time otherwise. He is also more adept at making better decisions, planning, and achieving goals.

Shower Often

I'm not saying that you shower daily, but you should wash your body and hair regularly on days that work out the best for you. Your body continuously sheds dead cells and skin that needs to be cleaned and taken off from your skin's top layer. If it accumulates for a longer time, it can cause severe illness and contagious diseases.

Cut Your Nails

Who doesn't like clean hands or feet? Keeping your fingernails and toenails trimmed at all times does not only attract people but also keeps you safe from problems such as dead cuticles, hangnails, and infectious nail beds. Adding to that, change your socks daily and *NEVER*, I repeat, *NEVER*, wear shoes without socks if you don't want to develop diseases, like Athlete's Foot.

Brush and Floss Daily

You should brush and floss your teeth after every meal, but since we're so occupied with life, it's recommended we brush at least twice a day. Brushing minimizes the risk of bacteria from accumulating in your mouth, which can easily lead to tooth decay and gum diseases. Flossing once a day helps maintain strong and healthy gums. In case you don't know, the bacteria and diseases that build up in your mouth can easily transfer to the heart and cause valve problems. Unhealthy gums can also lead you to lose teeth, making it difficult for you to chew and bite. To stay healthy and keep your smile alive, visit a dentist every six months for checkups and consultation.

Wash Hands

Incorporate washing hands in your daily lives so that it feels less of a chore and more of a habit that comes naturally. Wash hands after exiting the bathroom, before eating, before preparing a meal, after coughing, picking trash, and after shaking hands with strangers. You never know where those hands were before you shook them.

Sleep Well

Understand that 8-10 hours of sleep is not mandatory for everyone. Everyone has a different biological cycle, and it functions differently for everyone. Lack of sleep can make you feel low the entire day and make you feel groggy, thus making you compromise on your body, your health, and your immune system in particular.

Sometimes lack of healthy habits and hygiene can hint to something bigger than just poor living conditions. It could be a sign indicating depression; specifically, people who are sad or depressed neglect themselves to the point where they don't care about how they look, how they dress, or how clean they are. While talking about the importance of hygiene and its impact on our lives can help a few people, for others, it requires intense therapy and counseling to help them come out of their depression and sadness.

Effects of Positivity on Physical Life

"Wealth without health means nothing."

-Thabiso Daniel Monkoe, the Azanian

Psychologists are now discovering what most people had known for years - positive thinking and optimism have an impact on making people feel better. Positive thinking affects both physiological and physical health to a great extent. The same was discussed in Scheier, M.F., & Carver, C.S' research paper, *'Effects of Optimism on Psychological and Physical Health.'*

Several studies in the research paper considered the possibility that optimism may be beneficial for improving physical health. It focused on a few college students over the end of their academic semester, the most stressful time for most students. The optimists in the study relatively showed better results not only academically, but also physically. Since they viewed everything from an optimistic perspective, they would eat healthily, perform better, attend all classes, and even had full attendance at the gym as compared to the pessimists. Another study revealed that optimists were less likely to develop heart diseases, according to the Q-waves observed on their EKGs result. Their bodies rarely released a hormone called AST. Both of these factors hint at heart attacks. The results thus suggest that optimists were less likely to be infected during operation.

They also showed better results at recovery when compared to pessimists. Optimists were more likely to resume exercise and return to work full time. In addition, optimists regularly took vitamins and were least expected to eat unhealthy meals - meals containing a high amount of carbs and fats. There were also more chances of optimists getting themselves enrolled in a cardiac rehabilitation program. Thus, they participated more in physical activities than pessimists. We finally conclude that staying physically healthy can aid you in staying emotionally healthy, too. If you're taking care of your body and hygiene, eating the right food, exercising, and sleeping on time, you can cope with stress and fight sickness in a better way. Staying healthy provides a confidence boost not only to your mood - thanks to a clear mind, extra energy, and peace of mind - but also to how you look. Imagine feeling empowered because you know you are in control of how you feel, how you look, and how you carry on with your daily work. Overall, it's a fantastic feeling that comes with staying healthy and fits not only in your 20s, but also later till you're in your 70s.

Once you start noticing small changes in your life, given a changed lifestyle and improved habits, you will automatically be more mindful about your surroundings, what you eat, what you wear, what you drink, and what energy you attract. You will notice yourself prospering in all phases of life and learn to value the important things in life, such as family, relationships, health, and time.

CHAPTER:3

Chapter 3
Time

"If time were to take on human form, would she be your taskmaster or freedom fighter?"

-Richie Norton

What is the time?

The earliest time calculations were an observation of cycles of the natural world, using the arrays of change from day to night and season to season. More accurate time-keeping such as sundials and machine-driven clocks came along to place time in a more appropriate frame.

But what exactly are we trying to measure? Does time physically exist, or is it just a myth? At first, the answer seems clear – of course, time is real. It frequently reveals itself all around us, and it's hard to fathom the universe without it. Our knowledge of time became more complex as the world grew older, all thanks to Einstein. In his theory of relativity, he explained how time passes for everybody, but it doesn't always pass at the same frequency for people in different situations.

He resolved the mystery of time by aligning it with space to explain space-time, which behaves continually. His theory seemed to prove that time is threaded into the fabric of the universe. However, it failed to answer a significant question: how is it possible that we can move through space in any direction, but through time in only one? See, the universe is designed to move in one direction through time. Thinking of moving backward in time is simply unimaginable and beyond human knowledge.

One wonders, could time be some sort of impression created by the limitations of the way we view the universe? We're not sure of it yet, but perhaps that's the wrong way if we've been thinking all this time. Instead of questioning if time is a fundamental asset, it could exist as a promising one.

However, that has yet to be proven by scientists discovering this. This is on top of the many other time-related questions that remain unanswered to the human brain. If there is one thing that all psychologists, physicists, and scientists are sure about, it is that time is equal for everyone, whether a Muslim or a Christian, African American or a Caucasian, a child, or a retiree. Religion, race, age, and gender don't have an impact on time.

All of us share time irrespective of who we are. The only thing that differentiates us is how this time is managed. If you aren't making the most of the time in your personal or professional life, it is a grave mistake. Wasting time about your health may lead to unwanted results.

Doing the same in your professional life will prevent you from achieving your career goals. Stop wasting your entire time scrolling through social media, and use some of this precious time to accomplish those aspired life goals. Time is the most crucial currency because once it's gone, it cannot be reclaimed.

So, use this time to achieve your desired goals before it is too late. If you aim to be more productive at work or improve your relationships, this is a first good step in the right direction. You are verbally telling yourself to do better, and this marks the beginning of change. However, this intention to do better won't bear fruits until you take action. So, start using your time and take affirmative steps to fulfill your life.

Here's How You Can Manage Time Effectively

"How we spend our days is, of course, how we spend our lives."

-Annie Dillard

In the spring of 1997, NASA's Pathfinder spaceship landed on Mars and began transmitting unbelievable iconic images back to earth. After several days, something went dreadfully wrong. The process of transmission suddenly stopped. The spaceship was procrastinating: keeping itself fully engaged but failing to do its most important task. What happened here? There was a virus, it turned out, in its schedule.

Every operating machine has a program called the scheduler, sending orders to the CPU on how long to work on each task before moving on to another, as well as specifically what to move on to. If done right, computers function smoothly between switching responsibilities. This gives an idea that even computers get exhausted sometimes. Perhaps learning about the computer science of managing time can provide us with some idea about our human struggles with time management. One of the first understanding is that all the time you spend prioritizing your work is the time you aren't spending doing it. For example, let's say, when you check your email at work, you skim through all the messages, and then choose the first one you should reply to. Once you've dealt with that, you repeat the process.

It seems reasonable, but there is an underlying problem. This is known as a quadratic-time equation. With a full inbox, the prioritizing process will take more time. This means you'll be doing more work by several times. The engineers of the operating system Linux experienced a similar problem in the summer of 2003. Linux would rate every single one of its work in order of priority, and sometimes expended more time rating tasks than actually doing them.

The operators' counterintuitive answer was to change this full-rating process with a restricted number of priority 'buckets.' The system, as a result, was less accurate about what to do next but more than made up for it by spending more time making advancements. So with your emails, always choosing to answer the most important one first could lead to a disaster. Coming to an inbox packed three times more than usual could take nine times longer to reply. You'd be better off responding in sequential order or even at random. Try it. Surprisingly, sometimes doing things in the right order is the secret to getting things done on time.Another concept that developed from computer time management has a lot to do with one of the most prominent factors of modern life - interruptions. When a machine moves from one task to another, it has to run a process called a context switch, fixing its place in one task, and clearing out old data out of its memory to make a place for new data. Each of these actions comes at a cost.

The acumen here is that there's an ultimate compromise between efficiency and responsiveness. Getting the maximum amount of work done means minimizing context switches, but being approachable means responding anytime a new task comes up. These two ideologies are genuinely under pressure. Identifying this pressure allows us to decide where we want to attain that balance we desperately need. The most prominent solution is to minimize interruptions and distractions. The less evident one is to group them. If no notification or email involves a response more urgently than once an hour, say, then that's precisely how frequently you should check your inbox - not more than that. In computer studies, this concept goes by the name of interrupt coalescing. Instead of doing things as they come up, the machine assembles these interruptions together, depending on how long they can afford to wait.

Interrupt coalescing in 2013 inspired a massive development in laptop battery life. Prolonged interruptions allow a system to check everything at once, and then immediately re-enter a low-power state. This situation with a computer also implies to us that adopting a similar method may allow us to retrieve our own focus and send us to one of the things that feel so rare in today's hectic life - rest.

The reason why I demonstrated the association between machines and our daily life is because we are surrounded by distractions and interruptions that hinder our work, so much so that it becomes extremely difficult to manage our time effectively and pay attention to the things that require our complete dedication and commitment. To become more productive, most people keep themselves occupied in some way or another. They accept more work than they can realistically do, but that requires more energy as well.

As Henry David Thoreau puts it, *"It is not enough to be busy, the question is: What are we busy about?"*

It perfectly depicts our everyday lives, where we engulf ourselves in a pile of tasks but are unable to prioritize them and identify which job is more important. For us, time is more important than anything. We can't control it but should try to make the best use of it. With life getting out of control and ending in chaos, everyone wants more power over it. This is where time management steps in. Whether you're an entrepreneur, a student, or a housewife, you need to make a manageable to-do-list to be more productive. Effective time management skills are often the best way to enhance your productivity. Here are a few life-saving tips you can use to manage your time more productively and effectively.

Make a Schedule and Stick to It

A disorganized schedule only leads you to procrastination and a waste of your time. It leads to a decreased focus on your goals. It is recommended to keep track of your time by creating a schedule and going according to it. Make one for every hour of the day, from traveling to meetings and shopping. Organize your day based on your priorities. The more aware you are of where you spend your time, the more you will be able to hold yourself accountable for wasting time on irrelevant things.

Stay Organized

No one likes working in a mess. Imagine waking up to an unorganized room, or coming to an unorganized workplace or a messy kitchen; nothing can be more demotivating than this. It might sound absurd, but organizing your workplace, your kitchen, or your work area helps declutter, removes distractions, and increases productivity. This will also save a lot of your time as you don't have to keep searching for the things you need. Use some DIY tips to organize your workplace, home, kitchen, and desk. This is bound to save you a great deal of time.

Focus on the Right Things

As previously explained through the computer example, you don't necessarily have to spend too much time of your day on the things that can easily be done in one hour. Identify the things that are essential and can be done immediately. Sometimes, try picking tasks at random to see if that works best for you.

Track down Your Negative Habits

What if you spend time on getting a better idea of your negative habits? Be consistent, and be mindful of the habits you want to eliminate from your life. Also, spend some time tracking your unfocused work. There might be a chance that along the process, you discover that you spend most of your time on things that don't show any results. In fact, you can utilize your time on things that produce more results, like conquering a skill or getting a new degree. For professionals, time is money. The more you understand where your time goes, the more you'll be able to manage your time effectively. How you manage your time is in your hands.

Time Matters

"The bad news is time flies. The good news is you're the pilot."

-Michael Altshuler

Most people think of time management as a pre-determined schedule. That, in its essence, is true, but time management is a lot more than that. Time management advice that we often receive misses the crucial point. Apart from a schedule, you also need to manage your mind. It is almost impossible to stop yourself from dawdling or wasting time if you're not occupied with activities that are important to you. Organizing your time around our self is no easy task either. Ask yourself if you are on track? Are most of your time spent on the activities that move you closer to your goal? One time-tested idea is always to plan your day. This won't take more than 10 minutes, and you don't necessarily need a paper and pen for this. You can prepare a mental check-list of the tasks you need to get done throughout the day.

Victor Hugo best describes this, *"He who every morning plans the transaction of the day and follows out that plan, carries a thread that will guide him through the maze of the busiest life. But where no plan is laid, where the disposal of time is surrendered merely to the chance of incidence, chaos will soon reign."*

Tracking your time is another valuable method of managing your time. When you see things in black and white, you are inspired to change. For instance, if you put great value on physical fitness, yet you spend no time exercising, you'll clearly have no option but to rethink the way you are utilizing your time. Peter Bregman in the McKinsey Quarterly proved that it's best to determine approximately five priorities in each area of your life. 95% of your time should then revolve around these activities. He further recommends that you schedule your day revolving around your tasks by dividing a piece of paper into six sections.

As you go through your to-do-list, assign your tasks to these sections turn by turn. The sixth section should be the other 5%. These are low-priority tasks. Once you're done with the important tasks, you can come back later to this section of the paper. It's definitely not easy, and you will probably face difficulty in saying no to the activities that don't support your goals.

You Can Do It All, Just not at Once

Research has proved that it generally takes us about 10 thousand hours to master any skill set. Convert that into work hours. This means that you are required to work five years, five days a week, and eight hours a day to master your ability. In case, you have just started working toward your dream, and you are trying to juggle between family and a job, it is most likely to take you ten years before you become an expert in your profession. That, my friend, is a lot of time to waste on perfecting a skill that does not align with your real personality. On the other hand, if you taper your focus and finish that book you've wanted to write or the trip you've wanted to take, you'll succeed in those areas, and this will then will clear the path for your next goal. It takes self-correction to resist the pull of doing it all at once, or simply giving up as a whole when you realize you can't do it at all. In fact, the first phase is where most people give up.

This is exactly what I mean when I say to manage time. We should manage our minds first. Sure, you can earn a lot of money, experience, and success in the career you choose, but once you're done with your 10 thousand hours, would you not want to be engaged in an activity that internally moves you? Let's be honest: Many people spend their time doing things just for the sake of it, or they do it to stay busy. These trivial tasks are bound to make them numb to the reality that they are not using their greatest abilities. They are simply following a path laid out by a friend, a family member, or a counselor, instead of making the hard yet humble choice to line up their life with their unique talents and skills.

Try not to Be a Follower, Be Sure, Be Confident, Be You

Manage your time in a way that supports your dreams. This is the path less traveled by most of us. The heart of time management is self-management. If you are not hard-headed, it's easy to get knocked off the track by the push and pull forces external to you. You have to take out time to choose wisely. If you truly want to lead yourself toward your dreams, you must identify and work on your greatest strengths. You must follow your talent, instead of forcing it down while you waste away your time on the tasks allotted to you by others.

Make use of time right now. Ask yourself: what is that one thing you can do to move one step closer to your goal. Do you need to make a call? Write an email? Call a client? If yes, then do it. Make every minute count. Time management simply means managing your mind. It means that you don't need to pay heed to your internal critic or the whining child. Just listen to the strong voice inside of you, deep within your soul, that knows the next right step. Time comes with a high cost, and you only have one life. So ensure it works for you and not against you.

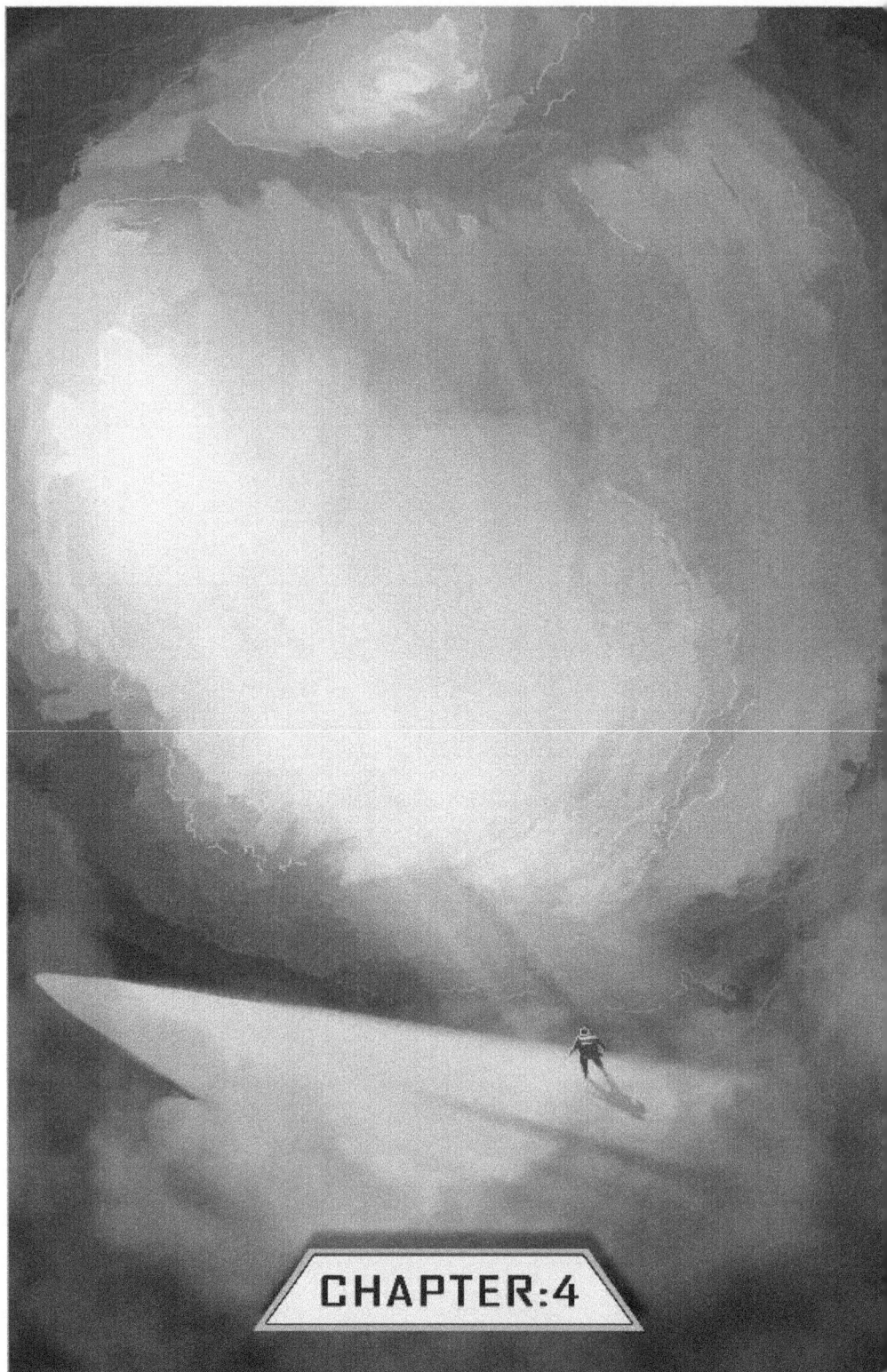

CHAPTER:4

Chapter 4
Opportunity

"Success is where preparation and opportunity meet."

-Bobby Unser

In 2003, a man was strolling down Boston Street when he came across a garbage truck running its daily course. The truck was standing at a pickup point, blocking traffic, with smoke and smell pouring out of its tailpipe. Despite its presence, the litter was all over the street.

There had to be a better way to collect trash and not pollute the air, the man, Jim Poss, thought to himself. Digging into the problem, he realized that these trucks consume more than one billion gallons of fuel in the United States alone. They run only 2.8 miles per gallon and are among the most costly vehicles to run.

In the early 2000s, metropolises and garbage collection services were considering more fuel-efficient vehicles and better collection routes to reduce their overall cost and environmental footprint. However, Jim was unconvinced that this was the right strategy. Through interactions with potential stakeholders, he overturned the problem and the solution. The solution might not be about creating a more well-organized collection process, but about the need for recurrent trash collection. As he pondered over this solution, he found its various benefits.

If waste containers could hold more trash, there won't be any need to empty them frequently. If trash did not need to be collected so often, collection expenses and the resulting pollution could be condensed. If litter bins did not overflow, there would automatically be less litter on the streets. Evidently, there were many more advantages to this method.

By utilizing solar technology at work, Jim anticipated how a new device might manage the trash better. His first idea of a solar-powered trash collector was terminated in favor of competing ideas for ecologically-friendly inventions, including the machine that would generate electricity from the movement of the ocean. However, his mind was still occupied by the problem and its various solutions. Jim started involving others, selecting a team based on the people he knew from his social network. He and his chosen team tried and tested multiple options and finally landed on The Big Belly – an invention that provided a perfect solution to the problems he pointed on the streets. This invention can hold up to five times more trash than conventional litter bins. As an effect, it radically increases the frequency of waste pickup and reduces fuel use and garbage truck emissions by up to 80 percent.

You see, people like Jim Poss create opportunities using simple practices that we observe on a daily basis in our lives. People, like him, are not big talkers. They just think and put their thoughts into action. I am sure all of us have come across a big talker at some point in our lives. The person who speaks significantly about what they will achieve, but at the end of the day, they don't do anything.

The reason behind their lack of success is the fact that they spend the majority of their time thinking rather than doing. If you want to be successful like Jim Poss, Bill Gates, Mark Zuckerberg, and a few others, take action now. Capitalize on every opportunity, adapt to the changes, and move a step closer to achieving your goals. The big talkers continue to procrastinate and daydream about their life goals. If you want to be a successful entrepreneur, search for an entrepreneurial opportunity and get out of a mental state of modern-day slavery, where you are chained to other people's expectations and are underpaid. It is time to be your own boss. Free yourself from the cumbersome burden of expectations and insufficient salary. Take advantage of every opportunity that is directed toward the entrepreneurial journey, and you will be on course towards freedom and independence.

Like most people, I have had a lot of good opportunities in my life, too. Some I took, some I missed. It is imperative to take advantage of as many opportunities as possible. You never know, because sometimes just like Jim Poss, one opportunity can mean the difference between an ordinary life and an extraordinary life.

Grab Every Opportunity

Sadly, opportunities do not last forever. The best part is that you can take the necessary steps to make sure you do not miss out on something truly amazing in life.

Learn to Say 'Yes'

Taking benefit of an opportunity in life begins with simply saying yes to them when they come knocking on your door. Remember, how badly you wanted to go on a trip but refused at the moment and later realized that you missed a great opportunity in life? That's exactly what happens when you turn your back to opportunities that come your way. Saying yes to yourself helps a great deal. Great opportunities often result in great ideas. If you are a pessimist about yourself, you will end up limiting your own options, and this is not a healthy state of mind to be in.

Don't Hesitate

Opportunities by their nature are short-lived. You need to be fast to make the most of them. For instance, many people hesitate to apply for a new job opening. They question themselves and are not sure if it's worthwhile. They simply wait for the right time to make a move. However, while they wait for the right time, someone else takes the opportunity. If you hesitate, you may stay behind and lose it.

Take More Risks

Oftentimes, opportunities, and risk-taking go hand in hand. To be honest, the best ones are often the riskiest. Someone who plans to start a new business is not only taking a significant risk but also taking advantage of an opportunity present to them. Ponder over the time when you were sitting on a couch watching television. How many opportunities came your way? Watching TV is less risky and offers hardly any opportunity. Sometimes, you have to kick-start those opportunities with a risk or two.

Maintain a Positive Attitude

Maintaining a positive attitude has multiple advantages. For starters, it helps you succeed when you are taking a risk. In the above example about a person starting their own business, how far do you think they will go if they have an extremely negative attitude while facing obstacles? Having a positive attitude about your odds of success can provide you with confidence.

Meet New People

When we talk about opportunities, it is mostly about who you know and not what you know. Imagine someone seeking a new job. One person has a network of four close friends and a few acquaintances, and another has 20 friends and 10 acquaintances who share the same industry as theirs. If their qualifications are the same, it is much more likely that the person who has a bigger network has higher chances of getting hired first. Why is that? It is simply because the person with more friends definitely has a bigger network. It is a fact that more jobs are found from networking than online job postings. So if you happen to know more people, opportunities will automatically follow you.

Be Inquisitive

Being inquisitive alerts and awakens your mind. Inquisitive people ask a lot of questions and continuously search for answers. Eventually, you will ask a question no else has ever asked before. Once that question is answered, you will come across a new idea. By asking a lot of questions about the world, nature, and almost anything, you develop a better understanding. Avoid being a passive player. Try and find out why things are the way they are. A lot of opportunities come and go without even you realizing it.

Focus

Make sure you know what you want in life. If you are clear of what you want in life, your mind will focus on that and will be in search of new opportunities to help you get where you want to. This helps when identifying the direction in your life. Someone with the chance to go to med school has a great chance, but if they dislike becoming a doctor, they should not avail it. The drawback for someone who would hate becoming a doctor and still goes to med school is that it takes away your time to do things you are more passionate about. This is what economists call an 'opportunity cost.' If this person goes to med school, they are going to spend a lot of time, money, and energy doing something they are not passionate about. This time, money, and energy could be better utilized in finding better opportunities they will enjoy more.

Stick to the Decisions You Make

The other side of being focused is having no focus at all. A lot of people do not have opportunities in their life because they lack decision-making abilities. In the above example with the med student, this person should stop from going to med school because they will hate it. However, if they are hesitant and unsure, they should go for it. It is always likely to make wrong decisions, but a wrong decision is still a better decision than no decision at all. When you are young, it is easy to get distracted and lost in the colors of life. These are the days when you should be more focused and determined to make the best of every opportunity. One of these opportunities that you should learn is how and where to invest. Of course, you can invest in your health, education, and life. But financial investment is one of the core activities that will benefit you in the long-term. You will not only save for retirement but also become more prudent and financially savvy. If you are still confused about why you should invest specifically in your early 20s, here are some reasons that will surprise you.

Start Building Your Retirement Account

If you have ever met a financial advisor to try to making sense of the numbers, you would realize how important it is to save a good amount of money for your retirement. If you begin now, you have nearly 40 years to plan, save, and invest, which will put you in a good position to retire contentedly when the time comes.

Benefit from Compound Interest

An escalating economy and rising inflation eat away at your spending power. The rising cost of living eventually makes daily provisions more expensive. This is why you should use the same principle of compounding to grow your investment. When you start investing early, you start compounding your returns as early as possible. This means you do not only earn interest on your initial investments but also continue to earn interest revenues on the interest returns you receive overtime. Doing this over the course of 40 years can make a massive difference between living a comfortable retirement and having to delay your retirement because you do not have enough money.

You Are No Longer a Victim of Market Volatility

When you invest in your 20s, you get an early head-start about the stock markets and property prices. This helps you understand how frequently the market fluctuates, and the prices vary. They sometimes fluctuate very dramatically, especially during the times of economic crisis or market booms. Of course, no one ever talks about having to deal with a thriving market. The problem appears when markets start to sink. If you are investing in the long-term, you can ride out the economic crisis and let your investment sit securely while waiting for an economic boost. When you start investing from a young age, you can strategize in advance to progressively move your investment from stocks or real estate to bonds. With time at your hand to plan, you can make a better financial choice and make use of new financial opportunities rather than panicking or stressing when you find yourself in a bad position later in life.

Avail Investment Opportunities

Life does not come with a guidebook. It is quite possible that in the beginning, you make some bad investment decisions as compared to when you are experienced and possess sound financial knowledge. This means that you will have a long time to continue earning and learning from those mistakes. If you are in your late 50s and start making investment mistakes, they could cost you a lot as you would have much larger savings and amount to play around with. Hence, you would not have a lot of time to recover from those mistakes.

Prepare to Deal with Future Expenses You've Not Thought of Yet

Imagine you are taking home thousands of dollars, and you are thinking of things you will spend that money on - maybe an extravagant wedding or purchasing an affordable home. The odds are, if you are just starting life, you cannot truly estimate the financial obligations and variations you will have to face over the next 40 years or when you will retire. Just to name a few – higher education, having children, nourishing, clothing, and educating them for the next 20 years, any likelihood of your parents needing some medical help, unemployment, severe illness, and a host of other uncertainties could take over your life.

The Bottom Line

Planning and saving for retirement is not the only reason to make well-planned investments. Many investments, such as those made in dividend stocks, can provide a flowing income throughout the life of the investment. Individuals in their 20s have certain benefits over those who wait to start investing when they are in their 40s and 50s. They are better learned, they understand market fluctuations, and are ready to take on newer opportunities as they follow.

Let's explain the benefits in two extreme cases:

Imagine there are two people, Emily and Rupert. Both start working at the age of 20 and plan to retire by 60. The market returns a seven percent interest compounded monthly.

Emily persistently invests $100 a month for 10 years. She stops funding when she turns 30 and leaves the money in the economy for the next 30 years until she turns 60.

Rupert waits ten years before he starts investing $100 a month into the stock market for the next thirty years until he is 60 years old.

Who do you think ends up with more money? Emily, who has diligently funded $12,000 or Rupert, who has personally added $36,000?

- **Emily:** $141,303.76

- **Rupert:** $122,708.75

Emily has added $24,000 less than Rupert, but since time is a close ally of compound interest, she takes over Rupert by a greater amount. If Emily had continued to invest $100 instead of stopping at 30, her retirement account would have looked a lot like this: $264,012.51. She made the best use of her opportunity to save and later enjoyed the fruits in her old age. So now you see, the power of saving and investing early in your 20s is substantial. So better start now than wait for the right time. However, that is all from the individual's point of view. What do you do when you are married and have children to look after? Are there any opportunities to avail that will benefit your children's future? Absolutely, yes!

As much as all parents love to invest in their child's well-being, education is an important aspect of a child's growth and learning process. While your children are young, it is an excellent opportunity for you to invest in their education. In the United States of America, a large number of children feel the harsh effects of poor health and academic failure. Most prominent among them is the family economic crisis, which is continuously related to negative outcomes. All across the U.S.A., 25 percent of children under the age of six live in extreme poverty and 12 percent in average poor conditions. Thanks to federal and state investments, the three most tremendous federal child care and early education programs have been implemented.

- Child Care and Development Block Grant (CCDBG)

- Temporary Assistance for Needy Families (TANF)

- Head Start

For child care and early education to be operative in promoting learning objectives for at-risk children, it needs to be of high quality. We all know how important education is in building a person's personality, their growth, and their future prospects. Availing the right opportunities that lead to these results is also equally important.

"Opportunity does not knock, it presents itself when you beat down the door."

-Kyle Chandler

Opportunities are a chance for you to try and turn hardships in your favor. You always have the choice to work on it happily and feel contended or carry an approach of procrastination to maybe just *'get it right the next time.'* However, when the next time does not come, all you have done is made a decision to linger on to agony, rage, and bitterness a little longer. Problems and hardships are guaranteed to come along, either before the opportunity occurs, or maybe right along with it. However, this should not be a reason for you to wait for some other opportunity to appear so you can take that instead. Remember your kindergarten school lesson about how many times opportunity taps on your door, and you will realize why. If this is the only chance you get, even if it has problems, don't you think you should take the risk on it?

If the universe is extra generous to you and throws an opportunity in your lap, I advise you to avail it. At least give it a chance, so that your future is not filled with *'What ifs'* or regrets of making a wrong decision without testing out the results. Allow yourself to nurture your abilities so that you can truly take care of your life.

To be precise, utilizing an opportunity to improve your life and your goals is a smart decision. People who cannot do that should stop whining and blaming external factors. You get extra credit if you learn from your experiences along the road. So go forward, grab that opportunity, step up to the trial, and bring purpose to your life.

CHAPTER:5

Chapter 5
Connect and Disconnect

"We need time to defuse, to contemplate. Just as in sleep our brains relax and give us dreams, so at some time in the day we need to disconnect, reconnect, and look around us."

-Laurie Colwin

Human beings are like animals. They are born with the need to connect with others, similar to them. Their capability to socialize urges them to disconnect from their personal isolation and connect with the environment external to them. Their need to satisfy their inner egos allows them to crave friendships and create positive interactions.

When you connect to your social media life, you automatically disconnect from reality, where you should cultivate friendships and develop interactions. As a result of this disconnection, communication skills worsen, and your relationships deteriorate. Sometimes, this deterioration could be preferable because certain relationships are negative and toxic. These relationships tie you with people who demotivate you and are draining. They bring you down and discourage you from taking the healthier steps in life. What needs to be done in this context? Remove the chains that drag you down into the abyss. If you lose a negative relationship, be glad because you would not be wasting time in negativity anymore. Contrarily, you can use this time to build positive relationships with people who encourage you to be a better version of yourself. Once you connect with other people in real life and build positive relationships, you will feel happier and productive. For a stronger and healthier relationship, you should always rely on good communication.

At different phases in your life, different things hold importance. In school and university, your priority is earning good grades or making new friends. Those math tests drain you, yet they matter the most. Once you graduate from university, getting a good job becomes more important. When you face failure or the first rejection, it is also important that you learn from your mistakes, which then takes you a long way ahead. Savings matter. Paying the bill on time matters. Financial steadiness matters. Investing in good relationships matters. Having a passion for constantly learning new things and experiencing new adventures matters. Seeking a purpose in life matters. You cannot avoid good health, either. On the other side of the world, to a poor mother, nothing matters more than feeding her hungry child. Now you understand the concept, right? As priorities change, the level of significance we place on things also changes, which subsequently transforms our perception of things that really matter in life. Through the trials of life, what is more important is how much love you have to offer. This may sound a little absurd, but at the end of the day, it all comes down to looking beyond yourself, your imperfections, your possessions, and your wealth.

Help others makes a positive change in their life. It works both ways if you do not expect selfish gains and returns in exchange. When you are generous to others, share with them, wave, smile, acknowledge them, spend time with them, and show any kind of love and kindness, it activates the areas of our brain connected with pleasure, social connection, and trust. Selfless behavior stimulates endorphins in the brain and improves happiness for the people we help and ourselves. Both sides equally benefit from one act of kindness.

Besides these standard affirmations, I also believe that helping people, connecting with them, and being compassionate give you a unique inner gratification that serves your soul in the long term. So, what needs to be done?

Be nice to people, take out time from the hectic monotonous life, and connect with them. The only thing you take to your grave is your flesh. What you leave behind is a whole lot of kindness, love, and good deeds that, in some way, have the potential to change the world for better. As hard as it may seem, change and loss are intertwined with life. We are unable to live our best life without both of them.

What really matters is to be present at the moment. Do not just exist. Live and make the most of that moment. Although it is difficult with so many distractions, in the end, they push you into a greater realization of what truly matters in life.

Why Is Building Relationship Important in This Digital Age?

"Nowadays we have so many things that take our attention - phones, Internet - and perhaps we need to disconnect from those and focus on the immediate world around us and the people that are actually present."

-Nicholas Hoult

Think about your closest friend and your favorite colleague. It could be your spouse or someone you went to school or college with. Irrespective of the person, know that your relationship was constructed over time by spending time together.

Our lives have been captivated by technology and ways to do things efficiently and cheaper, whatever suits us best. While I completely understand the need for financial stability and efficiency when needed, the area where you cannot achieve efficiency and speed is relationships and human connections. With the dawn of emails and WhatsApp, we have pushed relationship-building aside. Here are a few ways you can revive your relationships.

Spend Time Face-to-Face

To know someone better, you need to spend one-on-one time with them and gradually build the relationship. This also applies to business relationships, team members, vendor, and supplier relationships, and, most importantly, your customers. You cannot solely rely on emails, texts, blogs, and Facebook posts to add value to your relationship. I am not asserting that these are not crucial tools and do not have room in your life and business activities. However, they can never replace the need for face-to-face interaction.

The best investment you can make in your business and personal life is to spend time and create relationships that last long. Take time to get to know the people around you. Keep in mind that people are the common denominator in your business. They make, buy, sell, deliver, and use your products and services. So, you should never be willing to sacrifice the relationship by depending only on emails and texts for communication.

Understand Needs

It is in interaction and communication that you find people's real needs, wants, desires, passion, problems, and perspectives. Communication requires time and is not a one-time event. Make it your priority to step out of your technology bubble and build stronger relationships with people around you. It will pay big profits in the value of relationships and prospects to help and serve one another. Divert your eyes from your phone and focus on the person in front of you to learn more about them. Find ways to approach them and help them. They will find ways to help you, too.

Block Your Calendar Temporarily

Spare a few minutes and identify 20 people who can help you spread your business far and wide. These can be your friends, vendors, suppliers, employees, partners, or customers. Delete your calendar for the time being and focus on connecting with these people. Put them first on the list and start building your relationship with them. I suggest keeping one day of your busy week for disconnecting and connecting with people who genuinely add value to your life and your business.

Stay Committed

Building stronger and deeper relationships requires a commitment of time and effort. If you have trouble remembering, make a calendar, and stick to it. Remember, you cannot expect anything in return without giving something first. Invest time and effort in building relationships and watch how your life and business evolve in front of your eyes. It is more relaxing to have a cup of coffee with someone than terminating someone on an email and expecting a satisfactory response. Seek and focus on ways to develop relationships with people rather than technology and online strangers. Make it a priority to spend time with them. Not only will it help your business, but it will also bring joy to your life in general.

In a business setting, brands consider *interaction* to be another product to sell, but the communication is not a thing. The products and advertisements are how you communicate with your audience. It is a relationship. It is a way of delivering your message and your idea to the bigger audience.

Remember, All Relationships Are Based on Trust

You can put your trust in someone when you have confidence in their thoughts, sentiments, or talent. You feel that confidence because someone expressed their feelings or thoughts to you. In short, you trust someone because they exhibited consistency over days, weeks, and maybe months or years. Trust is simply a synonym for time, you can say. This is the reason why people feel connected to good managers and companies.

It is because the manager in the organization is reasonable in all situations, it is because the organization operates efficiently irrespective of the ups and downs of the market, and it is the consistency of their actions over the period of time that builds trust among its employees and customers.

This tells you a lot about relationships in particular. They require a lot of effort, time, and hard work.

Without Honesty, There Is No Trust

Trust and honesty go hand in hand. Of course, to trust someone, that person needs to be honest. You cannot trust someone if they are not honest about their thoughts, feelings, or intentions with you. Their dishonesty strips you of your ability to make good decisions. This is the reason why most people try to avoid managers who put on an act because the manager's lies and coerces employees into making poor career and work choices. This is why people avoid companies that do not provide right guidance because the organization's lies force investors to invest poorly or withdraw their funds.

Invest in Relationships, Not Things

The choice to invest in relationships instead of things has real outcomes for what you manufacture, who you ask to make it, and how you get the job done. If you want to initiate a conversation because what you value most about your customers is their knowledge, then there are appropriate methods, tools, and people to help you do that. Only a significant relationship with people can help you achieve your goal without any hindrance. Even your relationship can help you overcome these hindrances.

Why Is Building a Relationship So Complex?

In this age of technology, there is a propensity to perceive humans as machines. Do you recognize yourself as a machine? Do you consider that if you get the right inputs and have the proper system, you will function well without setbacks? Machines are practical and straightforward, but humans are more complex and social. They are not machines that require fixing. They are people who crave relationships with one another. They want to discover harmony and build unity.

Humans need encouragement to understand that they are good at survival. The choice they made in the past and continue to make because of their sufferings and experiences is what leads to their survival. Those same choices may also stop them from achieving personal success, secure connections, and happiness. To overcome a world of hurt that taught you how to live, you must be able to reach out to another - a better way of living. This can be achieved through hard work, efforts, and using the right tools.

The perfect tool is when a relationship is centered on kindness, honesty, self-responsibility, admiration, and love without judgment.

A great scientist and innovator of the 80s, Buckminster Fuller, once said, *"If you want to teach people a new way of thinking, don't bother trying to teach them. Instead, give them a tool, the use of which will lead to new ways of thinking."*

Research suggests fundamental pillars for any entrepreneur, organization, and person that can use them as a tool to improve their relationship with their external world.

Emotions Are Real, Not a Medical Issue

One does not have to be a psychologist to understand and deal with emotions. People who come from environments where they experienced sufferings are often stressed, depressed, and sad. Some people believe that in order to deal with negative emotions, one requires a degree in clinical psychology or social service. We need to understand that once a person is given space and assurance of their privacy, they will find ways to open up and face their sorrows.

Practice and Failure Require Empathy

It is only through the practice of failure that gives rise to learning. You cannot expect to walk into a gym and lose 10 pounds. That is not how the human body works. You lose 10 pounds by losing one, two, and three pounds initially. The body grows sturdier through practice and recurrence. However, we forget this simple rule when dealing with relationships in real life. If they dishearten us once, we apply a zero-tolerance policy. Instead, we should give time to relationships to grow and flourish.

A study conducted at the University of California, San Diego, suggested that when we observe others in painful situations, our visual cortex automatically releases hormones of empathy that helps us connect stronger. Their mistakes help us understand the lesson that could be derived from similar situations. And thus, you try your best to prevent them in your own life.

Do Not Let Online Relationships Affect Your Real Relationships

Relationships suffer from online interactions. It is much easier to harm friendships and people on the internet than in person because of the ease of creating misunderstanding by written communication. Be cautious about how you word every message you send, despite the context. Remember that every electronic message you send becomes a permanent part of your organization and your life.

Research conducted by McKenna, Green, and Smith in 2001 discussed how online relationships affect your relationships with people in real life. People who spent most of their time talking to friends online were rarely interested in things going on with their friends and family. Partners who were secretly involved in online dating did not even consider it was cheating until it was exposed.

Manage Your Time Well

Manage your time online with time spent with friends and family in real life. It may seem too palpable to mention, but it feels subjectively different to go out to eat with friends than to spend several days involved in back-and-forth text messages exchange. The latter drains too much communication and value. Our impact on one another is much more powerful when we meet in person. When a friend is going through a hard time, a kind smile or a warm hug has far more healing power than the craftiest emoticon to lift another person's mood.

Mistakes Teach You Lessons

A vow to growing, earning, adapting, and changing with time includes lessons from mistakes and failures. The more you learn from your mistakes, the more you can grow in your relationship. Healthy relationships, relying on the effort and practice of seeing the self as well the other person, are crucial. They are often hard to digest for those who have experienced early life traumas. The human need for healthy relationships is simple. However, the complicated nature of every human being is what makes relationships a challenge.

Thomas Merton, an American Trappist monk, wrote, *"We do not exist for ourselves alone, and it is only when we are fully convinced of this fact that we begin to love ourselves properly and thus also love others. What do I mean by loving ourselves properly? I mean, first of all, desiring to live, accepting life as a very great gift and a great good, not because of what it gives us, but because of what it enables us to give others."*

Therefore, connecting and disconnecting matter so much.

CHAPTER:6

Chapter 6
Simplicity and Focus

"Our life is frittered away by detail... simplify, simplify."

-Henry David Thoreau

We live in confusing times. We live in a time when complicated information blasts off from different channels and mediums. Due to this, our focus and attention waver constantly. This is also the reason this age is called *the age of distraction*. While humans have never been free from distraction – from answering phone calls to replying to emails – distraction has never been so occupying, so overwhelming, so obstinate, so pressurizing as it is now.

Phones are buzzing nonstop from emails and notifications from social media such as Twitter, Facebook, and Instagram, begging for your attention. The more connected you are to technology and its usage, the more you find yourself up to your neck in a flood of information. You are at the losing end in the fight for your attention and involved in a blinding act of multitasking. When you are doing a task, you have distractions coming your way from all directions. In front of you is the replication of the human brain – the computer with email notifications and a calendar reminding you of important meetings you have throughout your day. Then there is the enticing lure of the Internet, which offers an innumerable amount of information and exciting material that can drag you into the black hole of the web. Other things waiting for immediate discovery are the unlimited opportunities of discovering new things, both good and bad, for connecting with people, both strangers, and friends, for gossips, news, sharing a gaudy picture, and so much more.

During this time, several new emails arrive in your inbox, waiting for an immediate reply. Several tabs are open in your browser at once, each of them with the flashy user interface and fun activities found on various websites. Several friends are waiting for you to reply, sharing your attention even more.

This is all just in front of you. Don't forget the background distraction, a ringing phone, a loud co-worker, documents lying unattended, etc. With so many things fighting for your attention, and so little time to focus on the real task at hand, it's no surprise that you'll get anything done at all. It doesn't just stop here. When you go home from work, there is a pile of bills and letters that need your attention, a blaring television with 500 different channels, your ringing mobile phone, the delivery guy on the door, your kids, friends, roommates or spouse, and so much more that you can't ignore. The buzzing and the ringing and the vibrating just don't stop. This is disastrous and shocking. All of us have entered this age without any warning of what is happening or realizing the outcomes. Of course, you knew that the internet was flourishing. All of us were excited and curious about it.

We knew that mobile phones were becoming popular, smarter, and global, and whereas some people may have denied, yet others would have welcomed the innovation. The opportunities offered by this booming technology are good enough, the continuous distractions, the progressive pull on our attention, the pressure of multitasking, the disruption of our spare time, and our ability to live with an ounce of peace. However, maybe we did not realize how much this could transform our lives. Maybe some of you did, and perhaps some don't realize it even now. With so many things asking your attention, it's time you pay attention to this.

The secret is simple: focus. Your capacity to focus will enable you to produce in ways that maybe you haven't in a long time. It'll enable you to simplify things and find peace of mind. It'll allow you to pay attention to less important things and things that matter the most. While doing this, you'll learn to focus on the small things. This will change your relationship with the world and yourself.

It's not *something paradoxical like less being more*, but *less being better and enough*. Focusing on smaller things is bound to make you more effective. It will allow you to do less, and by doing that, you will have more spare time to do things that are more important to you. It will force you to choose and prioritize, and in doing so, you will prevent the extremes that have steered you in your economic crisis, individually and as a society.

Life can be complicated, or so you assert. However, this isn't really the case as you have made your life complex yourself. Overcomplicating your life when it is, in fact, simple can lead to dire results. You can regard simplicity as a form of focus. When you are standing in a room filled with various microorganisms, webs, strange swirling molecules, and a chair, focus on one thing, the room with a chair. Every bewildering thing doesn't matter, and you'll be free from all the issues regarding this odd surroundings. If you rely on simplicity, it will enable you to solve all your problems.

Similarly, businesses use simplicity to sell their products. Famously, Apple brings your attention to an iPhone's casing and user interface. In reality, we all know that it is a sophisticated device. Irrespective of which, the *simplicity* approach leads to aspired sales.

Simplify. Focus. Small things. Less is better. These are a few ideas that you will master in this chapter, and this will lead to better and bigger things in all aspects of your life. If you're someone committed to creating in any way and form, the focus should be your ultimate priority. The focus is important to all of you who create because, without focus, it is difficult to reach your goal.

Focus and Achieve Happiness

"Most of what we say and do is not essential. If you can eliminate it, you'll have more time and more tranquility. Ask yourself at every moment, 'Is this necessary?'"

-Marcus Aurelius

There is a lot more to focus on and interrupt than just creating innovative things. Often connectivity, distractions, and lack of focus can have an impact on your peace of mind, your stress level, and your happiness. Back when computers took only a small part of your life, there were times and ways to distance yourself quickly. Unfortunately, many people are still occupied at that time by watching television and playing video games, which is not really that productive. It's important to distance yourself from these constant distractions. These things stress your mind in ways you find it challenging to handle.

You need a peaceful time to reflect and envision, and some time for tranquility. Without it, information and feelings that you are unable to put to rest will often flood your mind. You need the rest.; your mind needs the rest; you need to declutter your mind and recharge your mental batteries. Quietness, tranquility, and solitude lead you to happiness when you make them a part of your life. What you can do during this time - read, nap, write, swim, listen, observe, play with your pet, study, think, disconnect and connect – is not as important as the simple truth of having that time to disconnect. We'll talk about how you can find the focus to do these things later in the chapter, but for now, you need to know how valuable these things are.

Focus Habits

"The law of harvest is to reap more than you sow. Sow an act, and you reap a habit. Sow a habit and you reap a character. Sow a character and you reap a destiny."

-James Allen

Maintaining focus and then creating is not just about disconnecting. You can simultaneously stay connected and focused if you develop the habit of blocking out every distraction and bringing your focus back to an important task. One of the simplest ways of doing that is by adopting some focus habits.

A habit is a set of actions you frequently repeat. You may have a habit of going to bed before 10 or a habit of saying your prayers first thing in the morning. One of the influencing things about habits is that we sometimes give them too much importance. They can be spiritual and mindful. When you become aware of them, you just rush through them carelessly. Carefully carrying out a habit is essential, particularly when it comes to focusing because many times, we get interrupted and distracted without realizing it. The distractions easily overcome you because you are not focusing on the task. So when you pay attention to your habits, it's much more natural to focus and do your work. A mindful focus on a habit prevents it from becoming a meaningless chore. It's crucial to give value to each habit so that you enable yourself to pay attention and not forget the habit when it's not suitable. For instance, you can start every habit with a breathing session to bring yourself to the present situation, declutter your mind of thoughts and irrelevant things, and completely focus on the habit itself.

Here are a few easy focus habits that will help you distance yourself from unnecessary distractions. It is a list of habits that I strongly recommend for every situation. Try them out and see which ones work the best for you.

Quiet Mornings

Start your day in silence before the chaos of the day infiltrates your peace of mind. If you live with your family, wake up before anyone else does. The secret to enjoying this habit is to focus on yourself. Try not to use the internet. You can turn on the laptop to write something meaningful. You can sit outside and drink coffee while reflecting and reading. You can start your day meditating and doing yoga or a quick run before you leave for work. You can quietly sit, play with your pet, and do nothing. The takeaway here is the peaceful time to rest your mind and maintain the focus you'll be needing for the rest of the day.

Begin Your Day

Begin your day by not checking your notifications or emails. Start it by making a to-do list of the things you need to do throughout the day. List your top three important tasks first, or if you like, just write one thing you wish to accomplish in the day. This will help you stay focus on what's really important.

Additionally, continue this focus habit by working immediately on the task on top of the list first. Repeat the process for a few days. Once you can master a single task each day, move on to several functions within a day while maintaining your focus altogether.

Refocus the Habit

While habits to begin your day are great, there are a lot more distractions that can get in your way to disrupt your focus. Refocus your habit after every hour or two. This should not take more than a minute or two. You can practice this by closing down unnecessary tabs and applications that hinder your productivity.

After every hour, get up from your seat and take a walk for 10 minutes to clear your head from negative thoughts, allowing the blood circulation to flow smoothly. Then, return to your list of essential tasks and get back on accomplishing them again. Only go back to your email and browser if you have mastered the habit of refocusing your habit. It is also recommended that you take deep breaths to bring back your attention.

Switch between Focus and Rest

This is similar to breaks during exercise. Switching between phases of tough exercise and rest works well because it enables you to do an intense routine, as long as you remember to rest between routines. Focus works the same way. If you give yourself enough time to relax, you can gain an enormous amount of attention. There are many routines to this. Some of them may include:

- 10 minutes of focus with 2 minutes of rest

- 25 minutes of focus with 5 minutes of rest

- 45 minutes of focus with 10 minutes of rest

To find the ideal time for yourself, you need to experiment to find the length and combination that works best for you. Some people prefer short eruptions, while others like longer phases of uninterrupted creativity.

Switch between 2 Focuses

Instead of switching between focus and rest, you can also switch between two alternative focuses. For instance, you can work on two different tasks at once or cook two separate meals at once. The key is not to switch to do it faster because there is a short time of regulation each time you move from one task to the other. You can work on a task for 10 minutes and then shift your focus on the other for the next 10 minutes. You can also stay focused on one task for as long as you want, and when you feel your focus losing, switch to the other task that seems more appealing. The best part about this habit is that alternating to a new task can give your brain some rest from the previous task and can keep you productive for a longer time before you are interrupted.

First Interact and Then Block

Set a watch and give yourself some time to respond to emails, Twitter, Facebook, conversations, and notifications like you would typically do. Once you're done, use an internet blocker to block all notifications and incoming distractions for a few hours while you focus on your task. Then take another short break to respond to messages and notifications or other small jobs you want to do, followed by another period of focus.

End the Day

At the end of every day, take a look at what you did and what can be worked on. Remind yourself to disconnect for the rest of the night to think about the things requiring your focus the next day. It's a good opportunity to reflect on your day and life in particular.

Weekly Focus Habits

Though it's not important to review the progress of your focus every week, it can be valuable to set aside 10 minutes every week to bring your professional and personal life back into focus. It is best to review your tasks to ensure you're not letting them out of control. Simplify your task list as much as possible. Review the habits to see what's working for you and what isn't. Simply reflect on what you're doing with focus and if anything needs to be changed in particular.

More Ideas

These habits are just a few ideas that most people find more natural to follow. You should find the one that is convenient for you. There are uncountable opportunities. Some ideas include sparing five minutes every hour to refocus, taking a short walk every hour to allow fresh air into your lungs, doing yoga or meditating at the start of every day, and exercising three to four times a week.

Give yourself the time you need to focus, disconnect from the complexities of the environment, focus on being productive, and practice breathing and other self-help techniques to sustain your focus.

Simplify

"As you simplify your life, the laws of the universe will be simpler; solitude will not be solitude, poverty will not be poverty, nor weakness weakness."

-Henry David Thoreau

Simplicity is liberation. Complexity is captivity. Simplicity brings happiness and peace to your life, while complexity brings anxiety, stress, and fear. One can see its real example given in the Bible: *"God made man simple; man's complex problems are of his own devising"* (Eccles. 7:30). Since most of you may experience true freedom through simplicity, you once again feel connected to simplicity as an inherent reality that results in an external lifestyle. Both the inherent and external parts of simplicity are crucial.

You betray yourself if you assume you can achieve the inherent reality without it having an intense effect on how you generally live your life. To try to live an external lifestyle of simplicity without the inherent reality leads to fatal legalism. Simplicity starts with an inherent focus and unification. Experiencing that inward reality frees you externally. Your words become more truthful and honest. The desire for status and wealth vanishes because you don't seek it anymore. You stop yourself from being extravagant. Not that you can't afford it, but you just don't value it anymore. You seek more experiences and soulful adventures instead of material things. You live in the age of technology and are affected by its broken and disjointed state. You feel trapped in a labyrinth of connections.

One minute you make decisions based on reason and logic, and the next minute out of fear of what the society would think of you. You lack focus on things that matter to you the most. Because your need for security has steered you into a crazy world of technology and attachments, it is far from reality. You need to understand that the desire for wealth and position in modern society is insane. It is madness because it has lost contact with the real world.

You crave things you don't need or enjoy. You purchase items because you want to impress people. Here, the usefulness of things exists and psychological needs take over instead. You are intentionally made to feel embarrassed to wear clothes or drive cars until they are trending. The media have convinced you that to be out of fashion is to lose touch with the real world. However, it is time you wake up to the reality that adherence to a troubled society is a sick mentality. Until you see how unbalanced your life has become, you are not ready to deal with the greediness inside you. The madness pervades even your society, your home, and your soul. Fearlessly, you need to communicate in new and more human ways to spend a life. You should stand up to the modern madness that defines people by how much they can earn or yield. You must experiment with new options for the present deadly and toxic system.

The Simplicity Circle

"Life is a series of natural and spontaneous changes. Don't resist them – that only creates sorrow. Let reality be a reality. Let things flow naturally forward in whatever way they like."

-Lao-Tzu

The simplicity cycle is a set of steps that apply to almost all situations. Each step uses a unique technique to help you declutter and keep moving forward successfully. It is equally important to plan and keep your future steps in sight because the decisions you make at the start of any task will show its impact down the road. Small strategies and reasonable thinking from the beginning can save you a lot of time and stress later on.

The Beginning

Begin by determining a few essentials and making a list of things you plan on adding to further steps. One way to implement this is by first creating an outline or a diagram that defines various parts or factors you're planning to use. It doesn't have to be chronological. You are just collecting some data at this point, and you can organize it later. The list doesn't have to be complete either. As you get started, you should expect your list to contain things that won't make it to the final step. For now, the aim is to simply devise components and aspects that will become a part of your final product. The point is to begin somewhere.

The Density Slope

Once you've structured the skeleton of your project or outline, start filling in the gaps that you feel need more work. This is where you start adding content, assembling the framework, and taking your project to the next level. This is also where you find out you may be ignored at a few points. It is a complicated process, but it's productive. Fight the urge to categorize, filter, and arrange things at this phase. Don't let yourself be distracted by organizing or editing. Don't second-guess, rephrase, edit, or delete anything. Let it stay. Your focus should be on creating things, figuring out the missing pieces, and adding them to the task.

The Shift Point

When your project hits the critical point of complexity, it's time to shift your focus. You've reached the point where you stop adding things and opt for a different set of tools, techniques, and strategies. It is quite easy to overlook many things at this point, and hence, continue to make amendments long after such amendments stop being productive.

This is very common when you're not focusing on, or you're not aware of the path that lies ahead. That is because the route of getting to this point generates inertia in your task. You've established habits and built up some speed that you need to keep track of. To make sure you're focusing on the right things, it is useful to look out for signs you easily identify throughout the process. They will be helpful to prudently stop and evaluate your own progress, checking to see whether you've hit the stage of complexity and need to shift your focus.

The Obstacle Slope

At this point, you need to keep in mind two points. First, identifying that maybe you're headed in the wrong direction. Second, doing something to stop that. The signs of extreme complexity you'll find in this step are similar to those you resolved in the previous one. However, the only difference is in the degree. The extra complexity at this stage could be coming from external factors that are out of your control. It could be rules, people, resources, or mental obstacles. To prevent them from getting in your way, you can take a pause, readjust your focus, and start all over again.

The Stop Point

It's understandable. You get excited, carried away, and distracted when things start going your way. But then, out of nowhere, things start becoming more complicated than ever before. Maybe your early modifications were not helpful, or perhaps, they were initially good, but your thought and anxiety ruined them. It's okay when this happens. It is not the end. When things go south, it calls for radical actions. It's time to implement new strategies and plan from the beginning. You can do this by first admitting that things are bad. Second, stop adding new things. Put down the pencil or shut down the computer and go for a walk to refresh your mind. Lastly, pick an eraser, delete that file, or simply start from the beginning.

The Simplification Slope

Some of the most useful techniques needed for this step are:

- Remove an irrelevant and unnecessary part of your work. In short, you may find yourself removing an unimportant or useless part, and find that your tasks still work fine without it.

- Blend multiple factors into a single component. Instead of two or more parts, they are arranged into a more systematic whole.

- Identify parts or factors that are not connected with your task, and you can do just fine without them.

Remember, the focus here is *simplification*. It functions best when you are mindful of the simplifications and keep them to a minimum.

The Ship Point

Great! You've finally arrived at the point where there is minimum complexity, and simplification is high. This is the ultimate goal. At this stage, the purpose of the project is highlighted and made simpler. You have arranged all the components required to achieve maximum functions and removed all the unnecessary clutter. The final picture is everything you wanted. Take a minute to appreciate the work that you have done. There is just one last thing left to do.

The Time Arrow

As much as you would like to bask in the glory of your masterpiece, the universe has other plans. And irrespective of how fantastic or simple your project is, time has its way of making it less valuable. This is simply because change is inevitable. Many external factors, including technology, economics, people, weather, opinions, and interest, may have adverse effects on your work. In short, time can move quality aside. However, you can overcome this phase easily by staying prepared for what could happen.

Write down a list of changes or external factors that hinder your productivity or stop your project from moving ahead. If you do this efficiently, you can quickly achieve focus and simplicity in all of your tasks without having to cease your work. With time, things are bound to change, but remember to stay positive, optimistic, and encouraging, and most of all, remember to lead by example.

CHAPTER:6

Chapter 7
Positive Attitude

"Positive thinking will let you do everything better than negative thinking will."

-Zig Ziglar

A positive attitude is the key to happiness. Positivity inspires you to adopt healthier habits, which lead to a healthy life. Your brain is built to identify threats so you can protect yourself. This survival instinct helped our ancestors avoid predators and invaders. However, we are not confronted with similar threats anymore. You may not need the survival instincts in the way your ancestors did, yet these are embedded in your brain even today.

Those survival instincts that once kicked in when wild animals threatened humans or other invaders are now activated after events such as heartbreak, professional failure, and other things like traffic jams or long lines at the grocery store. Hence, you find yourself with a stern attitude toward a life where positivity subsides. It is vital to train your brain to focus on the positive things that occur daily.

Regular practice of gratitude can accomplish this. Find the good things in your life, and be grateful for what you have. Soon enough, you will see a positive change in your attitude. Once you have this much-needed positive attitude, you will be on your way to happiness. Think of the time when you were unhappy. You probably associate it with a bad situation such as a tragedy, a problem you were going through, or maybe you did not have something you desired.

Maybe you mentally attributed the cause of your sadness to these external factors that were not in your control. People often think it is their situation, problems, and worries that make them unhappy, and that if their condition evolves to some extent, then happiness would find its way in their lives. Well, it turns out that it is not entirely your circumstances that are responsible for your changing emotions.

Truth be told, your wandering and negative thoughts are the real reason behind your unhappiness. People with perpetually lost minds are reportedly less happy than those who have the ability to focus on present tasks. History tells us that Buddhists, mentors, and saints long preached that a wandering mind gives room to unhappiness and dysfunction, and the secret to true happiness lies in mastering or having control over your mind, not in changing your external factors. The most shocking truth is that sadness does not just come from a negative mindset in relation to unpleasant things. Research shows that people with negative minds are less happy than those whose minds do not wonder at all. The research concluded that certain activities require the complete focus of the individual.

They were happier when their minds were fully present and at the moment. So, the verdict concludes that when the mind wanders frequently, it can adopt negative thoughts and drastically reduce your overall happiness and well-being. A negative one then replaces your positive attitude.

But, What Is a Positive Attitude?

Most of us already have an idea of what a positive mindset or attitude is. Let's see what the definitions tell us: *"Positive thinking is a mental and emotional attitude that focuses on the bright side of life and expects positive results."* Another well-explained definition of a positive mindset by Kendra Cherry is: *"Positive thinking actually means approaching life's challenges with a positive outlook. It does not necessarily mean avoiding or ignoring the bad things; instead, it involves making the most of the potentially bad situations, trying to see the best in other people, and viewing yourself and your abilities in a positive light."*

We can deduct from these definitions and come up with a reasonable definition of a *positive mindset* as the ability to focus on the right side, hope for positive results, and face challenges with a positive mindset. Possessing a positive attitude toward life means making positive thinking a crucial habit, constantly seeking the silver lining, and forming the best out of any situation you find yourself in.

Qualities of a Positive Mindset

"When life gives you lemons, you make lemonade."

This quote tells us that a positive attitude can make bad situations bearable. People who possess a positive outlook on life enjoy many other benefits, including a decreased risk of heart disorder and a longer life expectancy. A positive attitude can assist you in upholding the mental energy you need to work well in all areas of life. However, a positive attitude is not as simple as being happy at all times. It requires many helpful traits and powerful qualities, such as resilience and optimism. Now that we know what a positive mindset is and how it works, we can dig into the list of what a positive attitude truly looks like.

Acceptance: Realizing that things do not always work the way you want them to, and also learning from mistakes.

Optimism: The will to make an effort and take your shot, instead of believing that your hard work will not pay off.

Gratitude: Constantly appreciating the positivity in life.

Resilience: Getting back up from negativity, failure, and disappointment instead of giving up.

Mindfulness/Conscious Mind: Devoting your mind to conscious awareness and augmenting your ability to stay focused.

Integrity: The quality of being noble, righteous, and simple, instead of selfish and cunning.

These qualities do not only constitute a positive mindset, but they also work in other ways. Adapting to optimism actively, accepting things for what they are, being resilient and thankful, awareness of surroundings, and integrity in life will help you build and maintain a positive attitude throughout.

Become Successful through a Positive Attitude

"In every day, there are 1,440 minutes. That means we have 1,440 daily opportunities to make a positive impact."

-Les Brown

Since you have understood what a positive attitude is, you can turn to one of the biggest questions of all times: what purpose a positive attitude will serve? What is so important about having a positive attitude? Will I become successful with a positive mindset?

The qualities listed above should provide you with a hint. If you turn the pages in literature, you will find advantages associated with positivity, optimism, resilience, and awareness. You will observe that awareness and honor are connected to a better quality of life. Acceptance and thankfulness can lead you from an average life to an exceptionally amazing life.

The Significance of Having Positive Thoughts

Building a positive mindset and achieving those advantages is the work of the thoughts you nurture. Do not panic – this book is not a guide to positive thinking. I do not claim that by simply thinking happy thoughts, you will become successful in life. I, for sure, do not believe this kind of optimism is guaranteed in every circumstance, every hour of the day.

Giving a place to the right thoughts does not refer to being constantly happy or cheerful, and it is undoubtedly not about overlooking anything negative or hostile in your life. It emphasizes taking both the positives and negatives into your consideration and still choosing to remain optimistic. It is about realizing that you cannot always be happy, and that bad moods and emotions are a part of life.

They come when they are bound. You can do nothing about them, except welcome them and actively overcome them. Most of all, it is growing your power over your attitude in the face of whatever life throws at you. Of course, you cannot always control your mood and emotion, and you do not have power over the negative thoughts that pop in your head at any minute of the day. However, you can choose how to react to them. When you choose to surrender into negativity, cynicism, and sadness, you are not only surrendering to a loss of power and possibly indulging in adversity, but you are also missing out on an incredible possibility to grow and learn.

Psychologist Barbara Fredrickson says that negative thoughts and emotions have their place in your life. They enable you to refine your focus on threats, weaknesses, and dangers. This is significant for survival, maybe not as much as it was for our descendants. However, positive thinking and emotions widen and nurture your horizon, your resources, your skills, and simultaneously open doors for new opportunities.

Constructing a positive structure for your thoughts is not about being chirpy and annoyingly clingy, but investing in yourself, your well-being, and your future. Understand that it is okay to feel down or think negatively sometimes, but deciding on responding with positivity, resilience, and optimism will benefit you a lot more in the long run than most anything else. Implementing positive thinking at the workplace, at schools, and homes have huge advantages since it will reach a larger part of society and make a valuable impact. Managers can introduce positive thinking training among employees. At the same time, teachers can introduce fun activities that will not only engage the students but will also teach them a thing or two about a positive attitude.

Sometimes, the external factors cannot be controlled by the manager, teacher, and any other person. Here are a few strategies that are likely to be more effective and give you more control over implementing them.

1. Lead by example. Follow a positive, heartening attitude in all that you say, do, and believe.

2. Develop a positive learning space for your employees and students. Give them a chance to explore their potential.

3. Help your employees or students envision a positive result from every aspect before starting any task.

4. Eradicate negative thoughts from your employees' and students' thoughts. Respond to the common phrase that *I can't do it* by just saying it with, *Why*

*can't I do it? What's stopping me from doing it? How
can I help you?*

5. Assist your employees and students in transforming
    negative thought patterns into positive ones.

6. Act like you are your students' and employees'
    biggest supporter. Encourage them and boost their
    morale and self-confidence to expect the best results
    out of them.

7. Introduce an incentive or reward system to boost
    productivity at all times.

How You Can Develop a Positive Attitude toward Life

The secret to happiness, optimism, and successful life is
positive thinking. This does not mean they lack negative
thinking. Rather, they are seeking every possible way for the
positive. It is your choice to switch your mind from negative
thoughts to positive ones. You do not necessarily have to
wait for a bad thing to happen to you. You can start now
immediately if you want to live a happy and positive life.

1. Think of your mind as a meter that spikes your thought patterns as positive on one polar and negative on the other. Create an imaginary color for positive thoughts, maybe green. It strikes green when you have thoughts that increase, build, strengthen, and preserve your integrity. Think of the negative thoughts as the color red. It turns red when you have thoughts that drain your mental energy, absorb your goodness, and leave you feeling exhausted and pessimistic.

2. Train your brain to look for the good in everything. This is a way of turning on your intentional thinking. Train your mind to think positively instead of allowing it to wander and run on negative thoughts.

3. Be conscious of the direction of your thoughts. To do this, begin with your mouth. Every morning before leaving for your daily routine, repeat a set of positive affirmations that will help you remain on track throughout the day. Affirmations, such as *I am happy*, *I will be productive today*, *I will think of all the good things*, *I will embrace positivity*, etc., help. Remember, awareness leads to better and rational choices.

4. Connect to God. Devote some time to the worship of God and doing things that are acceptable and commanded in your religion. Getting to know your religion and offering prayers often help structure a positive mindset. It will not only increase positivity but also make you significantly happier and content altogether.

5. Start by evaluating your thoughts until you discover the good in them. You can do this by writing down your thoughts on paper and then examining them. If it is a negative thought, find out its source and eliminate it. Find out its origin and waive it completely. Even by examining it gives you the margin to change it forever.

6. Seek the lavish life. Your mind is currently searching for positivity and goodness. Even after the end of every misery, problem, or trial, there is progress to be garnered. If you trip on a stone, look under it for a lesson. This is your brain making wise choices. It does not avoid the pain of the present day, but it looks ahead to the benefits of the future.

7. Be careful of innate enemies of positivity. Power is an example to learn from. It clearly states *I deserve* and keeps you in an unhappy position. It expects its convenience over harmony. It exhausts you instead of relaxing your mind.

Realizing what you naturally do is essentially negative is the remedy to its suspension.

8. If you are spiritually strong, you can't help but start to look for the positive. God is the ultimate positive thinker. He integrates His thoughts into the man. Growing spiritually replaces your worldly rights with what is truly righteous. It reveals the negative by turning your vision into the positive. It is the ultimate source of your positive outlook on life.

9. Surround yourself with people you want to be like. Positive people develop fun, excitement, and interest in new things. New opportunities, thinking, and knowledge increases your potential as an individual. You will find yourself in a state of discovery while appreciating the normalcy of life.

10. Find positive role models, become friends with them, learn from them, and grow with them. They are the people who always remain positive regardless of the circumstances. They grow in the vessel God has given them. Find yours and embrace the things it has to offer you.

Positive Energy

"Be fanatically positive and militantly optimistic. If something is not to your liking, change your liking."

-Rick Steves

You do not have to talk to everyone, and yet you can send a positive vibe to others around you. Non-verbal interaction plays a vital role in communicating your thoughts to others. It is one way you can effectively transmit positive energy. The more positive you are, the easier it is for you to develop mutual trust and harmony in professional meetings or to build a better working environment for your employees. The key is to diffuse positive energy.

For example, you choose to initiate a conversation in a room full of strangers. Choose a particular stranger who seems welcoming, open for a conversation with you, and above all, positive. Positive people generally like to greet with a smile. They are relatively funny because they want others to enjoy their company. Positive energy is also expelled when you show an interest and take a step to introduce yourself. To master this, you should not interrupt your thought process and be patient for the outcomes.

You cannot only convey positive energy but can also create it. You can do so by simply saying, *"Thank you,"* *"Have a nice day,"* *"Sorry for the inconvenience,"* etc. Ultimately, society is expected to advance in general if we all start becoming nice to each other. We can do so by starting with these useful phrases.

To master positive energy, you should acknowledge what makes you attracted to positive people who are willing and excited to talk to you. Observe their body language. They will perhaps stand straight with their backs leaning because they are happy and comfortable to be present in the situation. Smile and see if they reciprocate. Imitate their attitude and implement it in your life.

We all experience positive energy once in a while. Visualize what it would be like to feel it every day, at school, at work, at home, and at any time. Here are ten everyday tips to elevate the positive energy in each day of your life.

1. Be comfortable with who you are. If you want to be different than others, you first need to learn how to *be.*

2. Make your mind stay positive at all times. Optimism is more fruitful than pessimism.

3. To achieve more, appreciate more. Less is more when you appreciate things for what they are. Happiness is not in a particular situation. Happiness seeks serenity, and serenity seeks peace.

4. Give up words like *awkward* and *weird*. We all are weird to somebody. People are not weird in general. They are just different in their own way. It is no surprise the word *weird* is often considered offensive and an insult.

5. The remedy to exhaustion is not to rest but devotion and wholesomeness. If you love doing what you do, continue doing and enjoying it more. Approach every step of life as a challenge that you have to overcome.

6. Pick your words carefully. Your words contain the power to heal or destroy someone.

7. Imagine where you want to be. Without a vision, you are stuck in the first chapter of life.

8. Do not stress too much. Stress is just a more significant problem that has not occurred yet.

9. Choose to transform yourself before life changes you due to circumstances.

10. Choose your view carefully. It paints your personality.

One Last Wisdom

If you are still connected to me after this very long read, thank you for sticking by. The purpose of this book was to help you find this publication worth the information you gathered from it. This book, a memoir of my experiences, my observations from both public and professional lives, is constructed to help the readers understand how the simplest things matter the most. How time, health, opportunity, and simplicity can enhance our lives and make them grander.

The gist of the book lies in how you can disconnect from material things and connect with the soul to find your purpose in true health and at the right moment. Carpe Diem! Seize the day to reveal your potential and let the course of life guide you to your real purpose. No matter where you live, what religion you follow, what God you believe in, your real focus should rely upon attaining simplicity, positivity, and tranquility in all periods.

The one last piece of wisdom from this book of hope, positivity, and optimism that I really hope sticks with you is this: Remember, positive thinking is a useful and powerful tool that can reward you and those surrounding you; however, nurturing positive thoughts all the time is very unrealistic and could even lead to disastrous outcomes.

You have numerous different emotions and thoughts, and you have a great selection of reasons for them, too. There are times when being a little negative can actually help you. In fact, it is a good idea to let out the negative emotions that you experience every once in a while. It is important to clear your system of these negative thoughts to make room for the positive ones.

If you are a positive person by nature, encourage gratitude for your intrinsic positivity, but make sure you do not push away the negative feelings that build up. They are an essential part of life, too. If you are a negative person by nature, do not desolate over ever thinking positively. Try out a few strategies that seem most convenient to you and give yourself a break if it starts to feel overwhelming. Do not forget that the goal is not to become an idealist, but to become the greatest version of yourself and maintain a healthy, happy, and positive lifestyle.

Ask yourself, how positive are you at the moment? Are you naturally optimistic or pessimistic? Are you aware of cultivating a positive attitude? If yes, then what are you waiting for? This is your moment. This is your chance to make a difference and embark on a journey unknown to you.

Thank you for reading this book, and good luck in cultivating a positive attitude.

Discover all the positive things

Life has in store for you

And move forward
To discover who you are
And fill your moments
With YES and POSSIBILITIES

Don't sit down with negative thoughts
Or bow to negative circumstance
Decide your future
And don't give up

Learn to stand up

To your ideas and dreams
And when you fall
Try to bounce back
And
Renew your spirit
And grab the moment
And prove your worth......

-Seema Chowdhury

AZIZ BENHAIDA